RAILWAY · TIMES ·

Introduction	3
7000 Not Out	4
Ashford's Ugly Duckling	5
Setting the Standards - BR Renewals	7
Doncaster Engineers Honoured	9
Highland Bonailie	11
Scrapping Resumes at Horley	13
Ivatt Atlantic Swansong	16
What's in a Name - The Fell Locomotive	19
Isle of Sheppey Closure	23
Royal Scot gets a Make-over	29
Branch Line Idyll	30
Last in First Out - A LIFO Closure in North Yorkshire	31
Regional Boundary Changes	33
Under BR's Umbrella	37
GNR Centenary Express	40
B1 Written Off	41
Ocean Terminal	42
Timetable Expansion in Size - and Price	46
The Red Dragon	47
Kerosene Castle	49
Around the Regions in 1950	53
Last Survivors	56
First SR Mainline Diesel	60
Lauder Branch Re-opens	63
Another Mythical Celtic Beast - The Welsh Dragon	65
Festival Preparations	67
Brief Encounters	71
Also making the news	76
In the next issue covering British Railways in 1951	80

RAILWAY TIMES 1950

Above : One of the graceful Ivatt Atlantics No 4403 here looking resplendent in its LNER apple green livery was photographed at Grantham station with a southbound express on an unknown date in the 1920s or 1930s. Dating from 1905 this example of the 4-4-2 class was based at Grantham during the 1920s and at New England during the 1930s. It was withdrawn in 1947 never receiving its new number of 2833. 1950 saw the final example of this famous class withdrawn from service. *(David P Williams)*

Front cover : The unique 4-8-4 Fell diesel mechanical locomotive No. 10100 emerged from Derby Works towards the end of 1950 with main line testing commencing early the following year. Although remaining in service until October 1958, apart from a lengthy spell out of action due to gearbox damage, BR appear to have lost interest in the concept being persuaded that the future ultimately lay with diesel electric traction.

Rear cover : The opening of the impressive Ocean Terminal in Southampton Docks on July 31 1950 was greeted with enthusiasm not only by the public but by the shipping companies and BR. However the glamour of the "golden age" of ocean liners was soon to pass away with the coming of cheaper air travel and by the late 1960s the two great "Queens" had made their final call. The QE2 was the last liner to use the terminal in 1980 after which the building lingered on until 1983 when it was demolished. *(Mike Ashworth)*

Copies of many of the images within **RAILWAY TIMES** are available for purchase / download.
Wherever possible illustrative material has been chosen dating from 1950 or is used to reflect the content described.

In addition the Transport Treasury Archive contains tens of thousands of other UK, Irish
and some European railway photographs.

© Jeffery Grayer. Images (unless credited otherwise) and design The Transport Treasury 2024

ISBN 978-1-913251-71-0

First Published in 2024 by Transport Treasury Publishing Ltd.,
16 Highworth Close, High Wycombe, HP13 7PJ

www.ttpublishing.co.uk or for editorial issues and contributions email to admin@ttpublishing.com

Printed in the Malta by the Gutenberg Press.

The copyright holders hereby give notice that all rights to this work are reserved.
Aside from brief passages for the purpose of review, no part of this work may be reproduced,
copied by electronic or other means, or otherwise stored in any information storage and
retrieval system without written permission from the Publisher.

This includes the illustrations herein which shall remain the copyright of the respective copyright holder.

Introduction

Volume 3 of "Railway Times" covers the year 1950 and features some of the main stories evident in the railway landscape during the twelvemonth. As well as scouring this first year of the new decade for stories of note I have tried to even up the BR regional coverage where possible such that in this issue we feature articles emanating from the WR – two new Welsh dragons, a closure in Pembrokeshire, and the gas turbine GT1. On the SR we have Ashford's ugly new diesel shunter, an Isle of Sheppey closure, the latest in the ongoing Leader saga, Southampton's new Ocean Terminal building and 10201 the region's first mainline diesel. The LMR is featured with stories about a Royal Scot makeover, Crewe's 7000th. locomotive together with the Fell experimental design. The ER and NER get a look in with the swansong of the Ivatt Atlantics, the honouring of Doncaster engineers through locomotive naming and an early closure in North Yorkshire. Finally the ScR is represented by features commemorating the withdrawal of two famous locomotive classes, the "Clans" and the "Lochs" and by the re-opening of the Lauder branch to freight following damage sustained in the severe 1948 floods. We also include details of a fascinating "might have been" relating to the independently operated Liverpool Overhead Railway.

Items of general interest concern the adjustment of regional boundaries, the new standard cab layout, changes in the shape (and price) of timetables, BR's renewals programme and the outline proposals for the new Standard designs resulting from the Interchange trials of 1948. As usual we aim to cover some of the quirkier happenings of the day in the "Brief Encounters" section such as the three hour whistling marathon inflicted on the residents of North Devon by a Bulleid Pacific and the multilingual announcements now available at Parkeston Quay. All articles are illustrated with images of the day, or as close as we can achieve, given that the Transport Treasury archive, although vast, does not cover every single aspect at the date we might wish to illustrate. I hope you find something of interest within and don't forget to look out for Issue 4 covering 1951 due to be published in the autumn.

Editor: Jeffery Grayer

A strange looking beast - No. 29951 one of the LMS diesel railbuses is seen here at St. Rollox depot on 31 May 1950. One of a class of 3 units built by Leyland Motors, its body styling revealing its bus origins, and introduced in 1933 they could accommodate 40 passengers but were withdrawn by BR in April 1951, having proved unpopular upon their transfer to Scotland where they became rather run down. Other snapshots of 1950 are included in the feature "Around the Regions in 1950" to be found on page 53. *(Neville Stead)*

7000 Not Out

A milestone was reached at Crewe Works this September when on the 15th. of the month it turned out the 7,000th locomotive to be built there. Suitably adorned with a tankside plaque reading "THIS IS THE SEVEN THOUSANDTH LOCOMOTIVE BUILT AT CREWE WORKS SEPTEMBER 1950", a humble example of Ivatt's 2-6-2 tank design No. 41272 was the recipient. Way back in February 1845 the first locomotive to be constructed at Crewe, 2-2-2 No. 49 *Columbine*, was completed there for the Grand Junction Railway and can now be seen in the Science Museum in London.

(Ed: No 41272 went on to be allocated initially to Bedford remaining there until transfer to Neasden in December 1958. Travelling to Exmouth Junction in June 1961 it remained in the West Country for two years. In July 1963 it was on the move again, this time to Leamington before a move to Llandudno Junction in November 1964 prior to its final move, this time to Skipton in August 1965 from where it would be withdrawn just three months later. It was scrapped at T.W. Ward of Beighton Sheffield in early 1966 (after a service life of just over 15 years).

The plaque fitted to the side tanks on Ivatt tank No. 41272 marking this milestone. *(Henry Cartwright)*

Left. No 41272 is seen during its two year sojourn in the West Country at Wadebridge in July 1962 with a service for Bodmin North. *(Don Matthews)*

Below. On 21 August 1954 No. 41272 is seen near Bedford shed, its initial allocation in 1950, where it remained until re-allocation to Neasden eight years later. It can be seen that the plaque was fitted just below the BR emblem. *(Eric Sawford)*

Ashford's Ugly Duckling
A new diesel shunter No 11001 for the Southern Region

Prior to nationalisation the Southern Railway had experimented with diesel shunters including Maunsell's three 350hp diesel electric shunters built at Ashford which had entered service in 1937. Bulleid's batch of 26 new Class 12 diesel electric shunters, featured in the previous issue of Railway Times, were being outshopped at the time this further design was being developed.

Designed by Bulleid, although not completed until after his retirement from the Southern Region, this 0-6-0 diesel shunter was powered by a Paxman V12 engine capable of delivering 500 bhp driven via a Vulcan-Sinclair fluid coupling to a synchro-self-shifting gearbox. This provided three forward and reverse gears in either high or low range, with speeds ranging from 5 mph in low gear up to a top speed of 45mph. Rather unexpectedly the locomotive's controls were laid out as for a steam locomotive, perhaps due to the fact that at that time there were few drivers with any experience of driving diesel engines. 4' 6" Bulleid-Firth-Brown type wheels were provided and the plate sided chassis was similar to a standard tank engine. The coupling rods, of differing cross section, were split as the drive from the rear jack crank shaft was taken to the centre axle through rather chunky connecting rods necessitating the provision of massive balance weights on the centre driving wheels and jackshaft cranks which did nothing for the aesthetic appearance of the locomotive. The body styling was unusual with a long bonnet housing a fuel tank at the front, rather than a radiator and a very high arc cab roof with large angled front windows. No. 11001 was originally finished in plain black livery but was later repainted green.

Its main duties were confined to branch line work and shunting, it being often noted on the Caterham branch during the 1950s being conveniently allocated to Norwood Junction Shed. It was also loaned to Feltham shed for a while before venturing further afield on loan to Stourton shed in Leeds in 1952. It was also based at Hornsey shed for a few months in 1956 before returning to Norwood Junction. However, it was only fitted with air brakes with no provision for train braking. This compromise meant that it was not entirely successful in performing either role, lasting less than 10 years in service. It was withdrawn in August 1959 and scrapped at Ashford Works in December of that year thus the ugly duckling never turned into a graceful swan.

Captured in the rather unlikely location of Derby shed in about 1953 No. 11001 presents a rather forlorn appearance. It is carrying a 20B shedcode which indicates Stourton mpd having spent two spells on loan working from this Leeds shed. It also visited Feltham and Hornsey sheds during its short career. *(J. T. Clewley)*

RAILWAY TIMES 1950

Shortly before withdrawal this Ashford Works yard view of 25 October 1958 illustrates how the whole gearbox / jackshaft assembly has been dropped out just like an axle set. Also clearly seen are the double brake shoes supplied for each wheel. *(R. C. Riley)*

This rear view was also taken at Ashford Shed earlier in the year on 21 June 1958 *(Alec Swain)*

Setting the Standards - BR Renewals

The first tranche of what would eventually be 999 new locomotives to a standard design was announced in 1950 with the first deliveries occurring the following year. This first batch, comprising 159 examples spread over six of the twelve different classes identified as being required, ranged from the Britannia Pacifics to the 82XXX tank locomotives. The remaining six classes missing from this first tranche were the sole 8P Pacific which was not built until 1954, the Class 3 77XXX 2-6-0s also not built until 1954, the class 2 78XXX 2-6-0 which came out in 1953, the class 2 84XXX tank locomotives, another 1953 introduction and the 9F 2-10-0s introduced from 1954. The table overleaf summarises this first tranche and sets it against the final total of Standard types.

Construction was spread over four locations, Derby, Brighton, Crewe and Swindon with works at other locations handling construction of and repairs to existing types. All of the Standards were designed to be mixed traffic locomotives thus providing maximum availability within their power range. Following the Interchange Trials of 1948, the performance of the various participants and the best practice of the various regions was studied to ensure that the new designs afforded maximum ease of access for servicing and repair, with the accent on interchangeability of parts, and an eye to economy of coal consumption. Increased mileage between repairs leading to economy of operation consistent with adequate reserves of power had been the goal of these new designs.

The new Britannia Pacifics would be capable of handling duties such as those currently performed by the West Country, V2, Castle and LMR 6P 4-6-0 classes. The other Pacific class with the lighter axle loading, later to be named after Scottish Clans, would similarly be capable of matching County, Black 5, B1 and LMR 5XP performance. The new Class 5 4-6-0s would be the equal of several current designs of that configuration whilst the lighter 4-6-0s, to be numbered in the 75XXX range, would replace the ranks of the vanishing 4-4-0 types. The new 2-6-4T design would be used mainly for passenger services whilst the 2-6-2Ts would handle shunting, freight and the lighter passenger duties.

The new order in the shape of No. 70000 *Britannia* departing Liverpool Street with the down "Norfolkman" on 29 September 1951 some nine months after delivery from Crewe Works in January of this new Stratford based Pacific. *(Roy Vincent)*

RAILWAY TIMES 1950

Turned out from Derby in April 1951 this is the doyen of the class No. 73000 *(Neville Stead Collection)*

BR RENEWALS PROGRAMME FOR 1951 – STANDARD DESIGNS

CONSTRUCTION	No. ORDERED/ FINAL No. BUILT	No. SERIES	ALLOCATION	WHEEL ARRANGEMENT POWER CLASSIFICATION
CREWE	25 / 55	70000 – 70024	ER 15 WR 10	4-6-2 7P6F
CREWE	10 / 10	72000 – 72009	ScR 10	4-6-2 6P6F
DERBY	30 / 172	73000 – 73029	ScR 5 LMR 25	4-6-0 5
SWINDON	20 / 80	75000 – 75019	WR 10 LMR 10	4-6-0 4
DERBY BRIGHTON	10) 44) / 155	80000 – 80009 80010 – 80053	ScR 21 SR 10 NER 3 LM2 20	2-6-4T 4
SWINDON	20 / 45	82000 – 82019	SR 10 WR 10	2-6-2T 2

Doncaster Engineers Honoured through Class A1 Naming

At a ceremony at Doncaster on July 13 three Class A1 Pacifics were named in honour of distinguished locomotive engineers of the past. The unveiling ceremony was performed by the son of one of those so named, Mr H. G. Ivatt the current CME of the London Midland Region. He recalled in his speech that Archibald Sturrock (1850-1868) was responsible for building Doncaster Plant, being one of the first to appreciate that large boilers providing ample steam were a prerequisite for successful forms of motive power. His successor, Patrick Stirling (1866-1895), would always be remembered for his famous "Singles" so vital for securing the GNR's reputation for speed and he made mention finally of his father H.A. Ivatt particularly famed for his Atlantics, the final examples of which were sadly to be withdrawn during 1950.

Neville Hill shed plays host to Class A1 No. 60118 named *Archibald Sturrock* in 1950 at Doncaster. *(George Barlow)*

RAILWAY TIMES 1950

Gaining admiring glances from a bunch of platform spotters No. 60119 *Patrick Stirling* passes through York non-stop with the "Flying Scotsman" in this undated view. *(Neville Stead Collection)*

No. 60123 *H.A. Ivatt* is captured at rest on Doncaster shed. *(Neville Stead Collection)*

Highland Bonailie*
Two notable Scottish Region locomotive classes have been withdrawn
* a farewell or parting toast

GATHERING IN OF THE CLANS

Posed by Inverness shed's triumphal pillared arched ornate water tower in this undated view is the last survivor of the Clan class No. 54767 *Clan Mackinnon*. The arch, which provided the gateway to the turntable and roundhouse, survived until demolition in 1963. *(Neville Stead Collection)*

The last example of the Cummings Highland Railway "Highland Clan" class 4-6-0 No. 54767 *Clan Mackinnon* was withdrawn in February 1950. The design dated from 1919 and only eight were ever constructed with the majority spending most of their lives at Aviemore depot, although the last pair to remain in service and the only examples to come into BR ownership had been based at Inverness depot since nationalisation. No. 54767 arrived at Balornock shed (St. Rollox) on February 4 1950 and left for scrapping at Kilmarnock Works under its own steam later the same day. Although it carried its BR number on the cabside the tender still proclaimed LMS.

DRAINING THE LOCHS
Another Scottish class to come to the end of the road was the Jones 4-4-0 "Loch" class when No. 14385 *Loch Tay*, the last example to remain in traffic, was withdrawn in April 1950. Given the BR No. 54385 this was never applied. Introduced in 1896 and at the time the largest passenger locomotives on the HR, eighteen examples were constructed by Dubs and by North British for the Highland Railway. After the Grouping they were rebuilt with Caledonian boilers. Some had been withdrawn as early as 1930 with just a couple coming into BR ownership working latterly from Forres depot.

LMS No. 14385 *Loch Tay* 4-4-0 of the Jones Loch class was captured on Forres shed on 14 June 1947. It would retain this number into BR days until withdrawal in April 1950. *(Neville Stead)*

Two years later No. 14385 was still based at Forres where it is seen on 28 July 1949. *(Sandy Murdoch)*

Scrapping Resumes at Horley - Past and Present

Towards the end of the 19th.c scrapping of LB&SCR locomotives was taking place at Horley in Surrey. The choice of this location, on the main Brighton to London line, was apparently dictated by a compromise. Horley was situated 25¾ miles from London Bridge and 24¾ miles from Brighton so pretty much spot on half way between the two. Apparently there had been some complaints from the purchasers of redundant steam locomotives, many of them London based, as far back as the 1880s that the carriage charges from Brighton to London were excessive and thus a compromise solution was found in that rolling stock would be delivered (F.O.C.) by the Brighton company to a point half way between.

One example of such a delivery was a locomotive built at Brighton Works to the design of J C Craven as LB&SCR No. 127. Originally a "West End" 2-4-0 well tank No. 127 had entered service in May 1858. As the design was not deemed to be a success No. 127 was rebuilt as a 2-4-0 saddle tank in December 1864. It went through a series of re-numberings beginning in March 1871 when it was allocated the number 106 this being followed in October 1876 by No. 292 and finally No. 367 in November 1877. Withdrawn from service 4 years later in December 1881 the locomotive was sold for scrap. This colourised image shows No. 367 waiting on the scrapline at Horley early in 1882. Apparently the purchaser was one Moss Isaacs. The details on the locomotive indicate that it was No. 2 in the batch sold to Moss Isaacs, below which is the weight in tons and hundredweights, 30 tons 8 cwt., and finally the date of sale 25/2/82. The image seems to have been reproduced commercially as a postcard with the tag line "A Relic of the Past". Moss Isaacs was born in the parish of St Judes in London in 1823 and by the 1851 census his occupation was given as "Dealer in Old Iron". From 1864 at least his business premises were advertised as being at 52 Bankside, Southwark, also called the Phoenix Iron Wharf and until 1878 he was in partnership with one Urban Gardner. Isaacs died in September 1889. *(Image courtesy Charlie Verrall)*

RAILWAY TIMES 1950

In 1883 Stroudley classified locomotives taken out of service into two categories – those on the surplus list which were generally stored at depots around the system and could be re-employed on such light duties as ballast workings and those lying derelict at Horley. These latter locomotives had their numberplates removed or the painted numerals crossed out, whilst the weight, date and scrap number were painted on for the scrapman's benefit and large white crosses, as seen in the image previous page, were painted on although this practice ceased after a few years. Locomotives often mouldered on at Horley for some lengthy periods and in 1885 for instance Moss Isaacs, who had bought the 2-4-0 illustrated above, purchased 13, including one that had been at Horley since 1881, for £2252 in May and another 27 for £4250 in September. Another key player in the scrapping process at this time was the firm of George Cohen who in 1880 cleared out a consignment of locomotives, tenders and old rails for the sum of £11,374, and who would go on to be an important player in scrapping BR's steam and diesel fleet during the 1960s - 80s. When sold removal of locomotives from Horley was often undertaken on a Sunday by a locomotive which had previously operated an excursion and had no booked return working. Delivery was often made to London's Deptford Wharf or Battersea Yard depending upon the purchaser's requirements. (Ed: Presumably charges were raised for the transfer of stock from Horley to the various sites in the capital but this would no doubt have been cheaper than paying for haulage all the way from Brighton).

Having set the background we may revert to 1950 when early in that year a correspondent to the *Railway Magazine* reported that "a new breaking up point has been observed at Horley". Details of the locomotives noted there awaiting scrapping during 1950 were recorded as –

Class	Nos.
K10	137, 144, 145, 344, 345, 386, 142, 151, 341
L11	167, 435, 168, 410
X6	657
G6	262, 275, 354, 240
E1	2097, 2127, 2162, 2112, 2690
T	500s
D1	2215, 2234*, 2274*
B4	2074
O1	1123, 1377
R1	1699
C2	2436

*These D1 tank locomotives had spent the previous 11½ years stored at Eastbourne before being towed to Horley in February 1950

Many of these withdrawn locomotives had spent some months previously on the scrapline at Eastleigh where the following, destined for Horley, were photographed in 1948 and 1949.

K10 No 344 at Eastleigh prior to removal to Horley *(Charlie Verrall)*

RAILWAY TIMES 1950

In a pretty parlous state 6X Class No. 657 awaits its fate when photographed on Eastleigh's scrapline on 11 August 1949. *(Leslie Freeman)*

Another destined for withdrawal in the not too distant future was photographed at Nine Elms where SR No. 386 of Class K10 is seen on an unrecorded date in 1948. This was one of several members of this class to be subsequently scrapped at the Horley site. *(Alec Ford)*

Ivatt Atlantic Swansong

The last example of the famous Ivatt Atlantics to remain in service, No. 62822, made its final journey on 26 November when it hauled a special service from Kings Cross to Doncaster. Withdrawals of the Class C1 had begun in 1943 and accelerated with post-war production of the Thompson B1 class. Seventeen C1s survived into BR ownership and although all seventeen were allocated BR numbers only No. 62822 received one. The C1 Class as it was known under both GNR & LNER classifications was designed by Henry Ivatt as an enlarged version of the C2 class and they were intended to haul the fastest and heaviest express trains on the Great Northern. Amongst the many passengers on board the valedictory special was the son of their designer H.A. Ivatt, Mr. H.G. Ivatt, who received one of the builder's plates from the locomotive as a memento. A circular headboard was carried occupying most of the smokebox and reading "Ivatt Atlantic Special 1902 -1950" and the locomotive itself put up a good performance touching 74mph with nine coaches in tow. The return fare, third class, was 19/3d and departure time was 11am with arrival at Doncaster at 2:15pm. BR made no secret of the

Above. The headboard carried by the special can now be seen at the NRM.

Left. The Ivatt Atlantic special awaits departure time at King's Cross. *(Online Transport Archive R509)*

ultimate fate of No. 62822 for on the poster advertising the trip they stated that it would be "Hauled by the last Ivatt Atlantic No. 62822 on her last journey to Doncaster, for breaking up."

On arrival at Doncaster a display was mounted consisting of the already preserved pioneer sister Atlantic ex GNR No. 251, withdrawn in 1947, together with a number of modern locomotives including A4 Pacific No. 60009 *Union of South Africa* and A1 No. 60147 *North Eastern*. The return trip to London was hauled by fellow A1 No. 60123 suitably named *H.A. Ivatt*. Whilst No. 62822 was subsequently scrapped at Doncaster a few examples of the class remained as stationary boilers at Doncaster and elsewhere and of course Ivatt Atlantics live on in the preserved pair No. 251 and the smaller C2 class GNR No. 990 *Henry Oakley*. The headboard from the Atlantic special can be seen at the NRM. (Ed: The surprise discovery of four C1 boilers, one of which had apparently belonged to No. 3287 which had been withdrawn in 1945, at a factory at Essex in 1986 led to the purchase of one of the boilers by the Bluebell Railway with the intention that it form an integral part of their exciting project to construct a new build LB&SCR Class H2 Atlantic. For as the LBSCR Atlantics were built to drawings modified from the GNR design this type of boiler is entirely correct for the new locomotive).

RAILWAY TIMES 1950

Last of the Ivatt Atlantics to remain in service was No. 62822 seen here at Grantham in its final year 1950. *(Neville Stead Collection)*

Wearing the fateful withdrawn symbol on its tender No. 62822 awaits the inevitable at Doncaster. *(Neville Stead)*

RAILWAY TIMES 1950

A4 Pacific No. 60009 *Union of South Africa* on display at Doncaster on 26 November 1960 marking the final runs of the Ivatt Atlantic. *(Online Transport Archive)*

Also on display at Doncaster on that day was A1 Pacific No. 60147, to be named *North Eastern*. *(Online Transport Archive)*

What's in a Name - The Fell Locomotive

Rather appropriately No. 10100 is seen passing the factory of Mirrlees Bickerton & Day at Hazel Grove Stockport, noted manufacturers of diesel engines. The company became part of the Brush group in 1949 and ultimately part of M.A.N.

So who exactly was this Fell character - the chap by whose name we have come to know No. 10100 the experimental diesel mechanical locomotive which entered the UK railway scene in 1950?

By the late 1940s the replacement of steam by diesel was perceived to be the way ahead but there was much debate as to which type of transmission future locomotives should employ. Should it be electric, as favoured in the USA, or hydraulic as preferred in Europe? A third option, mechanical transmission, while agreed to be extremely efficient, had generally been relegated to low-powered shunting locomotives or for light railcars, being generally considered to be unsuitable for higher powered operations. Enter Lieutenant-Colonel Louis Frederick Rudston Fell, a bluff Yorkshireman who fundamentally disagreed with this view and who had begun his career as a fitter and turner with the Great Northern Railway at Doncaster. He subsequently served with the Royal Flying Corps in the First World War where he had charge of an aircraft engine repair base, before he was elevated to become Head of Research at the Air Ministry and a member of the Royal Aeronautical Society. After joining Rolls-Royce he subsequently moved on to Armstrong-Siddeley and then reverted back to Rolls-Royce, where he became Chief Power Plant Designer and ultimately Technical Sales Manager. Whilst still employed at Rolls-Royce he formulated a concept of a railway locomotive that would do away with many of the perceived disadvantages of diesel traction. Theoretically his design would deliver a constant horsepower at the rail, irrespective of the speed of either the locomotive or of its power unit. To do this, his proposal involved the use of multiple engines feeding into a patent transmission. Concerns over locomotive weight were addressed by the use of several small, rather than one large, power unit which meant that both the engines and their supporting structure could be much lighter. This was also expected to save time in maintenance as a small individual diesel engine could be exchanged much more easily than would be the case with a larger power unit.

The design for 10100 was a collaborative venture between Fell Developments Ltd., Davey Paxman & Co., Shell Refining and Marketing Co. and H.G. Ivatt of the LMS. The 4-8-4 locomotive had six diesel engines, four of which were used for traction producing 2000hp through a single gearbox leaving the two auxiliary engines, both of which were 150 hp AEC 6 cylinder units, to drive the pressure chargers for the main engines. This arrangement was designed to enable the main engines to deliver very high torque at low crankshaft speed. The locomotive was assembled very quickly with construction taking place at Derby Works, being outshopped in November 1950. Early the following year main line testing began but in the first three months the locomotive only managed to cover some 100 miles primarily due to problems with the reversing dog. After resolution of this fault it was despatched to Marylebone station where it formed part of BR's display of modern traction in May which also featured the new Standard Class 4-6-0 No. 75000. The Fell locomotive was also exhibited at Eastbourne goods yard on 2nd June on the occasion of the opening of the Conference of International Railway Unions in the town. Along with 10100 were No. 70009 *Alfred the Great*, No. 75000, No. 73001, diesel shunter No. 15227 and "Booster" electric locomotive No. 20003. Also on hand were examples of Festival of Britain rolling stock such as dining and kitchen cars plus a variety of goods wagons and vans, semaphore and colour-light signals, bull-headed and flat-bottomed track in complete lengths. The Fell was towed to the site by No. 75000.

There followed a prolonged period of relatively successful testing on a regular St Pancras - Manchester diagram, the locomotive covering some 35,000 miles over the next year before considerable damage was caused to the gearbox by a loose bolt which had fallen through the gear train. This resulted in a twelvemonth mainline absence. Subsequent to this, BR appear to have lost interest in the project and development of an improved version was abandoned. However, No. 10100 did remain in service until 16 October 1958 when its steam heating boiler caught fire during a layover at Manchester's Central station. It was towed back to Derby Works only to be ignominiously ignored, BR being

RAILWAY TIMES 1950

Top left. Winning no prizes in a locomotive beauty competition the uncompromising nose of No. 10100 is to the fore in this view taken outside Derby Works.

Bottom left. The rather ungainly bulk of the design is evident in this three quarter front view also taken at Derby Works.

Below. Inside Derby Works the strange shapes of the Fell locomotive and one of the SR's diesels appears to elicit no interest from the little girl on the right.

persuaded by several manufacturers that diesel electric was the way ahead. It was slowly stripped of parts before being scrapped in July 1960. Fell died in 1977 having been awarded an OBE to add to his several military decorations.

(Ed: Shell produced a film about 10100 which can be viewed on Youtube and it explains in layman's terms the ingenious use of differentials, using the concept employed in a car's back axle to ensure the wheels turn at the appropriate speeds when cornering, which negated the use of a normal gearbox and allowed smooth acceleration from rest to top speed. The film makes the point that even the finest steam locomotives convert only 8% of their coal into pulling power whilst coal burning power stations that provide electricity for electric traction burn their coal at three times the efficiency of a steam locomotive. But the cost of electrification is great being only justified on busy routes that carry frequent services. With diesel-electric locomotives, 20% of the power is lost through the conversion to electric transmission which is necessary as a diesel engine by itself cannot start a heavy train from rest and accelerate it up to high speeds. It was expected that through this novel Fell transmission system only 6% of the power of the diesels would be lost).

RAILWAY TIMES 1950

Out on the road in 1950 the Fell garners more interest from several platform observers.

The Fell locomotive and diesel shunter on exhibition at Eastbourne. *(Charlie Verrall)*

Failure to Take Off - Isle of Sheppey Closure
Services between Leysdown and Queenborough have been withdrawn

The Sheppey Light Railway which was engineered by the redoubtable Col. Stephens and opened in August 1901 closed on and from 4 December 1950. Originally provided with five intermediate stations an additional two halts were opened in 1905. Operated by the SE&CR and latterly by the Southern Railway until nationalisation passenger traffic was always light so it was not long after opening that experiments, to effect economies of operation, were undertaken with petrol railcars. Unfortunately this was rather short lived as no local mechanic could be found to maintain these rather temperamental early vehicles. After a period of reversion to steam haulage, steam railmotors were used and remained in operation until WW1. Elements of these railmotors lived on, for although the engine components were scrapped, the carriage portions were paired together and coupled using a shared bogie and in this guise they remained in use on Sheppey hauled by regular steam locomotives up to closure of the line.

Apart from carrying local passengers, the line handled farm produce and a few tourists and it was also to become important in the history of early aviation. In 1909 the Royal Aero Club established their flying grounds at Leysdown and Short Brothers opened the first UK aircraft factory nearby at Shell Beach. Shortly after the whole operation moved to Eastchurch, the site being just to the south of Eastchurch station. In 1911 the Royal Naval Aviation School was established on part of the Aero Club's grounds and many famous aviators took their first flights from here, perhaps most well known being Winston Churchill in 1913. During WW1 a siding was built from Eastchurch Station leading directly into the aerodrome. RAF Eastchurch, as it became known, remained active during the inter-war years and was home to 266 Squadron during the Battle of Britain, finally closing in 1946.

When the railway line was first mooted, there were ambitious plans to develop a tourism industry on the island including a hotel at Leysdown and the provision of various camping sites. There was even a plan to construct a 7000 foot pier at Minster in order that pleasure steamers from London could call en route to the fleshpots of Margate. Sadly, like so many other grandiose plans for nascent tourist resorts, these plans never left the ground and the expected rise in passengers after the end of WW2 never materialised. Sunday services were cancelled and only four trains ran on the line each way on Mondays - Saturdays. The competing buses were popular however and such was the decline in rail passengers that gross receipts for 1949 totalled just £5,753, of which only a paltry £106 had come from passenger traffic. In the light of these figures and as the SR anticipated having to spend £28,000 on permanent way renewals in the future the decision to close was a foregone conclusion. The last train left Queenborough for Leysdown at 4:27pm on the afternoon of Saturday 2 December.

1950 view of locomotive running round at Leysdown terminus. *(Lens of Sutton, Denis Culum Collection)*

Articulated set No. 514 seen at the Leysdown terminus in 1950. *(Lens of Sutton)*

One of the halts was named Brambledown seen here in the last year of operation of the line. *(Lens of Sutton)*

The rather bleak terminus at Leysdown seen from the top of the water tank. *(Lens of Sutton)*

Top left. A London Chatham & Dover Railway chair seen attached to a rather rotten sleeper survives in the sidings at Leysdown in 1950. *(Lens of Sutton)*

Above. Another LC&DR survivor is this milepost seen alongside the track near a caravan site on the Sheppey Light Railway in 1950. *(Lens of Sutton)*

Bottom left. A third LC&DR relic on the line is this combined home and shunt signal. *(Lens of Sutton)*

Top right. The rural nature of much of the route is typified by this view taken at a crossing near Eastchurch. *(Lens of Sutton)*

Bottom right. One of the more important stations was that at Minster-on-Sea much of the settlement of which is seen in the background. *(Lens of Sutton)*

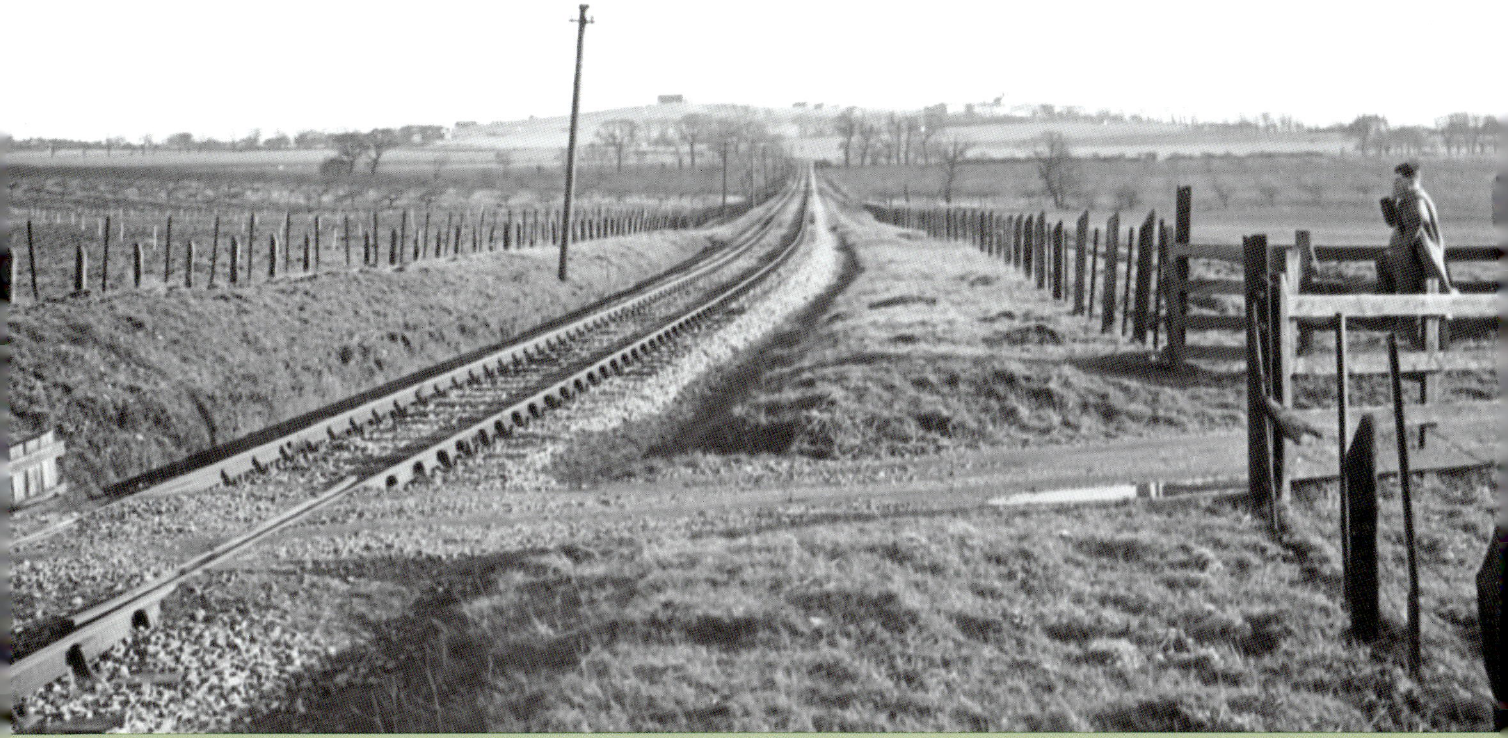

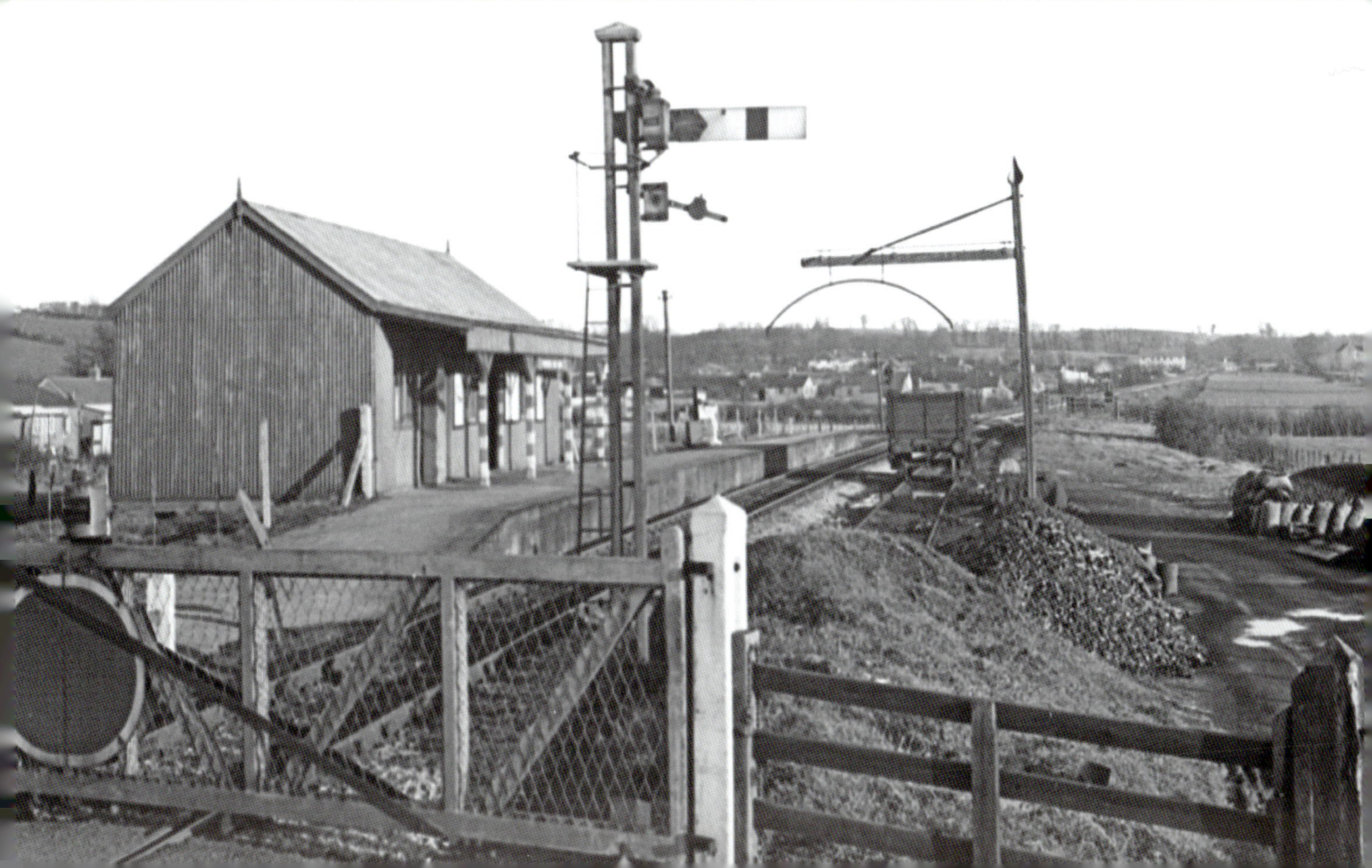

Above. Taken from the station footbridge at Queenborough, the stock for the Sheppey train is seen in the bay platform to the right. *(Lens of Sutton)*

Left. The penultimate stop before the junction at Queenborough was Sheerness East. *(Lens of Sutton)*

Bottom left. Ex SE&CR third brake coach No. 3561 is seen at Queenborough in 1947. *(Lens of Sutton)*

These coaches were originally built by the SECR as steam railmotors, but in the 1920s, they were converted into 2 pairs of pull-push coaches and 2 sets of non pull-push articulated coaches. They lasted in this state until the 1950s. One coach in each set had its luggage/guard section converted to passenger use, initially 1st class in the pull-push sets. This is one of the converted coaches, which was used in No. 514 articulated set, and had a luggage/guards compartment. Associated mainly with use on the Sheppey Light Railway, and afterwards on other lines including the West London, the other set was numbered 513.

Royal Scot gets a Makeover

1950 HEADBOARD

LATER VERSION WITH TARTAN BACKGROUND

A new headboard has been made for the "Royal Scot" express. The earliest headboard made of cast iron was introduced in 1933 and remained in use until the present time. A wooden version has been introduced by BR in 1950 as illustrated above which includes a shield depicting a Scottish lion (rampant). Later on a steel version (1953) and an aluminium version (early 1960's) would be introduced, both versions carrying the rampant lion and the name "Royal Scot". An aluminium version in the style of other named passenger trains was introduced by BR in 1951. In addition to the new headboard introduced in 1950, coach roofboards have been painted in the Royal Stuart tartan with the lettering "Royal Scot" or "London (Euston) – Glasgow". Menu cards in the restaurant car are distinguished by a "Royal Scot" emblem whilst dining car staff will wear a similar device in the lapels of their jackets. The entrance to platforms 12-15 at Euston has been enhanced by the erection of a sign incorporating a Scottish lion and this has also been done at Glasgow Central.

Left. The up "Royal Scot" waits to leave Glasgow Central for London Euston behind Camden Pacific *City of Coventry*.

46231, *Duchess of Atholl*, at Carlisle Station with the "Royal Scot" headboard, 20 May 1955 (R K Kirkland). *(Lens of Sutton Association)*

Branch Line Idyll

A rare colour glimpse from 1950 reveals a Petersfield service waiting departure time at Midhurst in May 1950 with 2 coach pull-push set No. 732 leading. At this date the set consisted of corridor third brake No. 2642 and corridor composite No. 4757. Between them they provided 64 seats third class and 12 seats first class – probably far more than the branch warranted and it was to be only another 5 years before complete closure of the Midhurst-Petersfield section came. Midhurst would continue to be served by freight trains from Pulborough until 1964. Further images of the Midhurst lines are contained in the Editor's recent book "Rails Along The Rother" published by the Transport Treasury. *(Roy Vincent)*

Last in First Out - A LIFO Closure in North Yorkshire

The Forge Valley Railway, a 16¼ mile single track branch line built by the North Eastern Railway serving a number of villages between Pickering and Scarborough, joined the York – Scarborough line at Seamer Junction. Opened on 1st May 1882, it proved to be one of the last railways to be constructed in North Yorkshire and indeed one of the first to close on 3rd June 1950. A short section from Pickering to Thornton Dale, which remained open for parcels and freight, was left in situ to cater for quarry traffic until 1962, following which the remaining track was lifted in January 1963. The branch, which had six intermediate stations, was single track throughout with a passing loop at Snainton.

Running close to the Pickering to Scarborough Road, now the A170, the line was vulnerable to competition from the United Automobile Services bus company and this was exacerbated by the fact that some of the stations were not conveniently sited for their respective villages. Despite economies employed by the use of steam railcars and pull-push trains over the years passenger numbers declined to the point where it was no longer viable to keep the trains running and services were withdrawn from 3 June 1950.

G5 Class 0-4-4T No. 67273 waits with a Pickering train at a very damp Snainton on 27 May 1950 just a few days before withdrawal of the passenger service. These tank locomotives were found widely over the North East working branch line passenger and some of the heavier suburban services. The class remained intact until 1950 when the first withdrawals occurred and although none survived into preservation one new member of the class is currently under construction by the Class G5 Locomotive Co. *(Online Transport Archive)*

One of the intermediate stations at Ebberston is seen here in this undated view. *(Neville Stead)*

By October 1962, the date of this view, the platform at Thornton Dale seems to be providing a home for some chickens. The end of quarry traffic here during this year would see the finish of freight working on this remaining section of the former branch. *(Neville Stead)*

Regional Boundary Changes

Following the establishment of the six BR regions in 1948 there had been some minor tinkering with boundaries at the start of the year with some more major reallocations towards the end of the year which would ultimately lead to the publication in 1950 of an 8 page booklet entitled "A Step Forward – Revision of Regional Boundaries". This detailed some more significant adjustments to the boundaries of the regions and therefore of the routes that lay within each region which would come into force from 2 April that year. The thinking behind these moves was prefaced in the booklet with a brief history of the development of railways in the UK at a time when there seemed almost no limit to the promotion of new lines, leading to competition in some areas where prospects appeared bright for more than one company. Amalgamations and voluntary working agreements between companies developed as time passed, such that by the time of WW1 the number of companies was down to around 120. Even with the Grouping, the waste of the old competitive days was not entirely overcome with overlapping and duplication of routes remaining in some parts of the country. With nationalisation in 1948 came further benefits of unification, one of which was the advantage to be derived from revising regional boundaries such that overlapping was eliminated, thus reducing administrative costs and simplifying supervision of operations. The aim was to arrange administration and supervision in such a way that, as far as was practicable, each region became self contained and the unnecessary expense of two or more administrations operating in the same territory was eliminated.

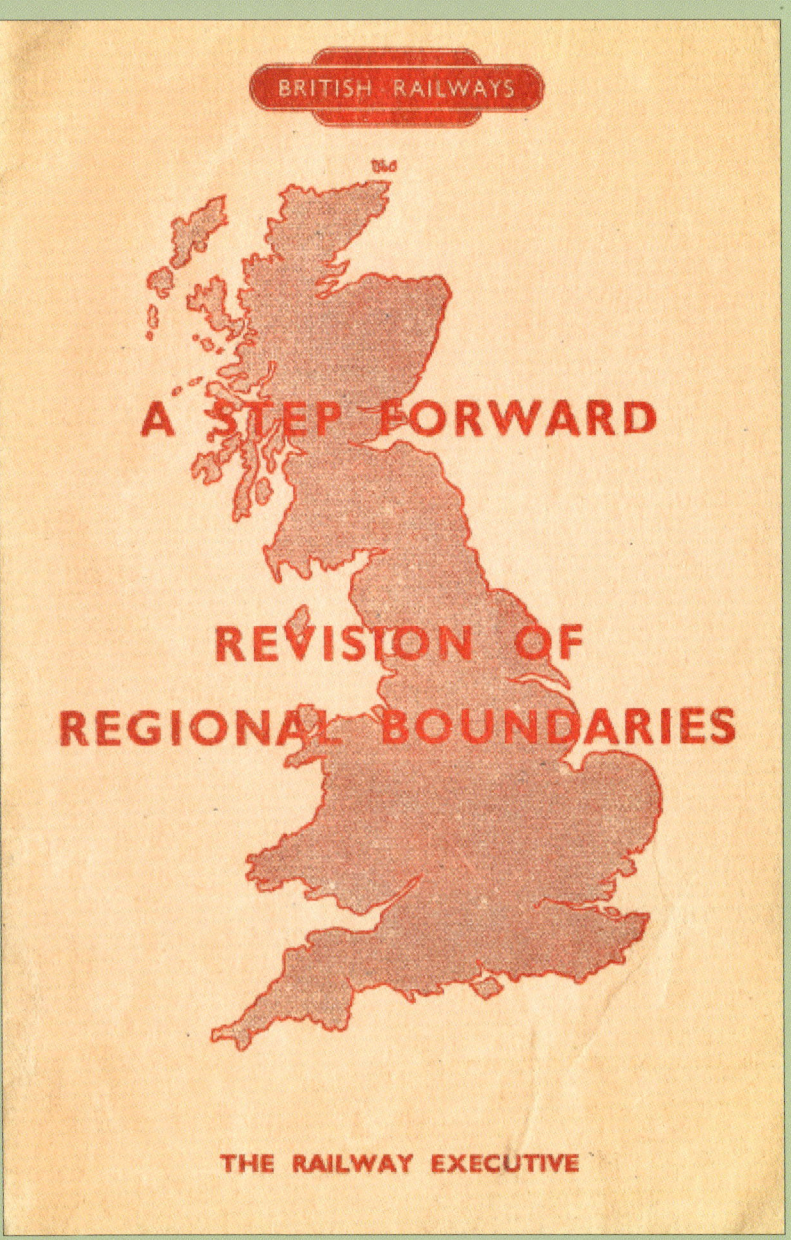

The public too would benefit from dealing with just one district commercial officer and one regional headquarters when in the past they may have had to deal with more than one. Benefits for staff also accrued, for instead of the present four Sectional Councils for each group of staff covered by the negotiating machinery, there would be just six, one for each region. Promotion opportunities would also be made easier as all job opportunities within a region would be available to all staff in that region and this would reduce the extent to which staff are obliged to move home in order to gain promotion, a definite advantage in those days of acute post war housing shortage. On those lines which saw heavy flows of traffic, the current operating arrangements for trains and traffic working would continue, for example the former LMS Birmingham and Bristol mainline would continue to be operated by the LM region, although it would come within the revised boundary of the Western Region. Similarly the former LNER (GC) route from Marylebone to Manchester London Road would continue to be operated throughout by the Eastern Region although parts of the line will fall within no less than three regions namely the Western, London Midland and Eastern.

The booklet concluded with the following aspiration

Kirkby Stephen East now transferred from the North Eastern to the London Midland Region. *(Neville Stead Collection)*

"Unification presents an opportunity to the railway industry to economise and to improve its efficiency, thus giving the public a better service. The revision of regional boundaries takes advantage of the opportunity. With good will and good work we shall take another step forward to that success which is the common aim of all who serve with British Railways."

Quite early on in 1948 a number of changes to the original boundaries were made which in summary encompassed the following :-

WESTERN REGION
Deleted – Ealing & Shepherds Bush (excluding Ealing Broadway station)
North Acton – Greenford electrified line (excluding Greenford station)
SOUTHERN REGION
Deleted – Turnham Green – Ravenscourt Park
St. Mary's Junction – Whitechapel Junction
LONDON MIDLAND REGION
Deleted – Carlisle Kingmoor mpd
The boundary between the LM region and the Sc region is now immediately south of Gretna
EASTERN REGION
Deleted – Finsbury Park – Drayton Park and Finsbury park LT Underground station
Electrified line through Stratford for LT services
East Finchley to High Barnet and Mill Hill East
Woodford to Leyton Junction (excluding Woodford station)
Fairlop loop (excluding lines south of Newbury Park)
NORTH EASTERN REGION
Deleted – Sprouston and Carham stations
Added – Silloth branch (Subsequently from June 12 1948 transferred to LM region)
Hexham – Saughtree
Reedsmouth – Morpeth
Rothbury branch
SCOTTISH REGION
No change other than redrawing of boundary at Gretna between LM region and Sc region

On 30th November 1948 an announcement was made by the Railway Executive that an extensive programme of inter regional adjustments was in hand to **"simplify supervision, to reduce administrative costs and to avoid unnecessary duplication"**. The routes affected were generally those that penetrated from one region into another and were a reflection of the historical competition that was rife during the period of "Railway Mania". The major re-allocated routes and individual stations that had been approved at that date were as follows –

LM region to E region
    Spalding Goods depot
    London, Tilbury & Southend section of the former LMS (Fenchurch Street – Shoeburyness, Barking – Tilbury and branches *

LM region to NE region
    Garsdale - Hawes

LM region to W region
    Craven Arms – Swansea Victoria and branches (Central Wales line)
    Swansea St. Thomas – Brynamman
    Merthyr Tydfil – Abergavenny and branches (Heads of the Valleys line)
    Hereford (Moorfield) – Three Cocks Junction
    Rotherwas Junction – Red Hill Junction (Hereford)
    Merthyr – Morlais Junction (formerly joint line)
    Former Joint stations including Abersychan & Talywain, Pontardulais etc.

E region to LM region
    Glazebrook – Wigan Central and St. Helens Central
    Bidston – Chester and Wrexham
    Former LNER Stafford route from Egginton Junction – Dove Junction
    Broomshall Junction – Stafford
    Hawkins Lane Goods depot (Burton)
    Oldham (Clegg Street) Goods depot
    Bidston Goods depot
    Birkenhead Dock Road and Duke Street Goods depots

NE region to LM region
    Carlisle – Silloth

W region to S region
    Newbury – Winchester Chesil (DN&S route)

Sc region to LM region
    Carlisle Canal mpd

Former Joint stations and depots now reallocated to one region
    LM region gains Tebay, Penrith and Normanton
    W region gains Bristol Temple Meads, Worcester Shrub Hill, Churchdown, Chelsea basin

* LM region retains Tottenham & Hampstead Joint line between Kentish Town and St. Ann's Road stations inclusive. S region takes over Gravesend landing stages.

Details of the remaining changes, not previously mentioned above, covered by the booklet "A Step Forward" were as follows –

The charming rural station of Cole on the Somerset & Dorset line which was now no longer in Southern territory but marked the southern boundary of the Western Region. *(Henry Priestly)*

NE region to LM region
- Kirkby Stephen East – Appleby and Clifton Moor
- Kirkby Stephen East – Tebay

LM region to E region
- Willington – Cambridge
- Raunds – Huntingdon
- Thorpe – Peterborough
- Wakerley – Wansford
- Morcott – Peterborough
- Edmonthorpe – Little Bytham
- Carlton & Netherfield – Lincoln Whitefriars
- Farnsfield – Rolleston
- Shirebrook West – Shireoaks
- All former LMS lines north and east of Hasland, Dore & Totley, Sheffield and Barnsley

LM region to NE region
- All former LMS lines north of Penistone, Darfield, Denaby and east of Skipton, Eastwood and Diggle

LM region to W region
- Hadley Junction – Coalport
- Bicester - Oxford
- Warwick Milverton station
- Leamington Spa Avenue station
- Banbury Merton Street station
- All LMS lines south west of Selly Oak to Bath and Bristol including branches
- Broom – Byfield (exclusive)

E region to LM region
- Mill Hill – Edgware
- Hill End – St. Albans London Road
- Harpenden East – Dunstable
- Former GC line from Quainton Road (exclusive) and Ashendon Junction – Heath (exclusive)- Marefield Junction – Leicester Belgrave Road
- All former LNER lines in the Nottingham and Derby areas west of Netherfield, Leen Valley, Bilsthorpe including Pleasley and Mansfield Central
- Hazlehead Bridge – Manchester London Road and to Manchester Central via Fairfield Junction and Levenshulme South
- Metropolitan & GC Joint line Rickmansworth (inclusive) to Quainton Road and Verney Junction

E region to LTE
- Metropolitan & GC Joint line from Harrow – Rickmansworth (exclusive) and Watford

E region to NE region
- Former GC & GN lines north of Barnsley Old Mill Lane – Mexborough – Doncaster

E region to W region
- Marylebone – Northolt Junction
- Neasden – Harrow (exclusive)

To E region
- Shaftholme Junction – Knottingley

S region to W region
- All former SR lines west of Exeter (Cowley Bridge Junction)
- Cole – Bath including branches to Burnham, Bridgwater and Wells

W region to LM region – see below
W region to S region – see below

The 1950 booklet set out these changes in tabular form and below is an example of one page which details all transfers from the Western region to the London Midland and Southern regions.

LIST OF MORE IMPORTANT LINES AND STATIONS INVOLVED—continued

Description of Line, Station, etc.	To be transferred From	To
Crudgington to Nantwich	Western Region	London Midland Region
Thornfalcon (Somerset) to Chard (Central)	Western Region	Southern Region
Thorney & Kingsbury Halt (Somerset) to Yeovil	,,	,,
Sparkford (Somerset) to Weymouth, including Bridport, Abbotsbury and Easton Branches (Dorset)	,,	,,
Grafton & Burbage (Wiltshire) to Andover Junction, including Tidworth Branch	,,	,,
Newbury (exclusive) to Winchester (Chesil)	,,	,,
Reading West (exclusive) to Basingstoke Branch	,,	,,
Westbury (exclusive) to Salisbury	,,	,,

Under BR's Umbrella?

In 1950 there were moves afoot to incorporate the Liverpool Overhead Railway (LOR), known locally as the "Dockers' Umbrella" and which had continued to be run independently following Nationalisation, into BR ownership. The adjacent dock was named after the Herculaneum Pottery Co. which had previously occupied the site and here 3 car set No. 44 runs into Herculaneum station in June 1951. Note the steam locomotive undertaking some shunting at ground level and the use of both semaphore and colour light signals on the LOR. *(Milepost Mace)*

The LOR was one of the select few small standard gauge railway undertakings, along with the likes of the Corringham Light, Derwent Valley Light, Easingwold, North Sunderland and Swansea & Mumbles that were not incorporated into the nationalised undertaking in 1948. Other exclusions were of course a number of narrow gauge, colliery and industrial lines which also remained outside the nationalised fold. By far the most important of the lines excluded was the LOR which carried a significant number of passengers and which continued to be operated by the LOR Co. The line did not fall under government control during the recent war and the company paid a dividend of 2% on ordinary shares the previous year.

The National Union of Railwaymen (NUR), no doubt fully aware that the "writing was on the wall" over the long term future of the line, made representations to the Ministry of Transport (MoT) suggesting that the LOR should be absorbed into the London Midland Region of BR. The British Transport Commission (BTC) although not having the power to compulsorily purchase such railways as the LOR did have the power to acquire it through voluntary agreement. The Railway Executive debated the issue, not just in relation to the LOR, but to those other lines which had not yet been nationalised. (Ed: By 1950 BR had already taken over the Easingwold Railway and closed it as recalled in Railway Times No.1).

A brief word about the LOR. Opened in 1893 with lightweight electric multiple units, it originally operated over a 5 mile stretch which was later extended at both ends to run the 7 miles from Dingle to Seaforth & Litherland. At its peak some 20m journey were made annually but the railway suffered greatly during the Liverpool Blitz and from the effects of corrosion caused by steam locomotives serving the docks passing underneath the ironwork of the overhead structure. Much reconstruction work was necessary in post war years and, although some cars had been modernised in the late 1940s, many required replacement with some being more than half a century old. (Ed: BR decided not to incorporate the LOR which was probably just as well because a report produced in 1955 into the condition of the line would show that many elevated sections required major repairs estimated to cost £2m (£67m at today's prices) over the following 5 years. The local company had insufficient financial resources to fund this work and the railway closed at the end of 1956, despite many public protests with the line being dismantled the following year. Although BR was not interested in taking over the LOR it did however flirt with it on occasion for, in an echo of pre Grouping arrangements with the L&YR, between 1947-1956 it allowed Aintree Grand National race day specials to operate over BR tracks from Seaforth & Litherland to Aintree using LOR stock. This involved the 500v electric stock of the LOR operating in series mode over 3 miles of BR track energised at 630v. Some arcing was apparently experienced!)

RAILWAY TIMES 1950

The guard of Car No. 1 smiles broadly en route to Seaforth from Dingle in this undated image. *(Arthur Mace)* *(Advertisement courtesy of Mike Ashworth)*

SEE THE DOX!

When in LIVERPOOL do not fail to see the WONDERFUL DOCKS and GIGANTIC LINERS

Permits to view LINERS issued at Booking Offices, PIER HEAD or JAMES ST. STATIONS

ROUND TRIP FARE
9D. THIRD CLASS
1/- FIRST CLASS
13 MILES

Special Arrangements made for large parties.

Apply—
MANAGER,
31 James St.,
LIVERPOOL

CHILDREN UNDER 16 HALF PRICE

RAILWAY TIMES 1950

The train arriving at Seaforth Sands from Dingle in June 1956 is just about to pass the junction signalbox. *(W. A. C. Smith)*

Set No. 16 sporting an early version of "speed whiskers" arrives at Pier Head station with a service from Seaforth in June 1956 the final year of operation. *(W. A. C. Smith),*

GNR Centenary Express

Above. The GNR centenary express awaits departure time at King's Cross with No. 60113 of New England shed at the helm which was described at the time as being in "spotless blue livery". *(Meredith 130-7)*

Right. No 60113 seen soon after its arrival on York mpd. The rather stern portrait of Denison is apparent and I'm sure he would not have been amused at the demolition which occurred on the special's return journey. *(Meredith 130-10)*

A special train was run to commemorate the centenary of the Great Northern Railway on 16 July 1950. Starting from King's Cross and hauled appropriately by A1/1 Class No. 60113 *Great Northern* which was a Thompson rebuild of a former class A1 constructed in 1922. The load of 305 tons comprised eight coaches consisting of six Pullmans and two kitchen cars. The large headboard carried a somewhat fearsome image of one of the founding fathers of the GNR, namely Edmund Denison who was chairman from soon after its founding until 1868. His obituary described him in rather unflattering terms thus – "brusque in his manner, impatient to a degree of human vanity in all its ugly shapes, and with little trace of sentiment or poetry of any description". The outward route of the special was via Peterborough, Boston, Lincoln and Retford to York. The return journey was apparently rather more eventful with the locomotive demolishing a set of crossing gates near Selby. Some passengers managed to salvage some of the debris and at the Peterborough stop the driver was apparently asked to autograph pieces of the splintered wood!

B1 Written Off
Accident near Witham 17 March 1950

At 3.11am on the morning of 17 March 1950 a collision occurred in fog near Witham Junction on the Colchester main line of the ER of BR. The 11.0 p.m. up express passenger mail train running from Peterborough to Liverpool Street via Ipswich over-ran all signals at Rivenhall and collided at about 60mph with the 7:45am mineral freight train from Whitemoor yard to Witham which was approaching Witham Junction outer home signal at slow speed preparatory to entering the up loop. The goods guard and the fireman of the express were both killed. There were only about 20 passengers and postal staff on the train and seven of them received minor injuries. The driver and second guard of the express were also injured, the former seriously. The locomotive, B1 class 4-6-0 No. 61057, and the first four vehicles of the express were badly damaged with the following two coaches becoming derailed but the last four vehicles remained on the track. The brakevan and last seven vehicles of the freight train were demolished, and 14 more were damaged. About 100 yards of the up track were destroyed, both lines were blocked by debris and both signal and telephone communications between Witham Junction and Rivenhall were cut. The wreckage was cleared rapidly and the down line was reopened by 6.0 p.m. the same day followed by the up line shortly afterwards. The official report into the accident concluded that the most probable cause was the fact that the driver of the express was travelling too fast for the proper observance of signals under the prevailing weather conditions. The signalman was also held partly to blame for not promptly calling out fogmen and for not instituting double block working following the passage of the freight train until the fogmen were in position.

Above. Class B1 No. 61057 in happier days seen here at Ipswich on 22 August 1948 some 3½ years before it was to be written off in an accident. *(Neville Stead Collection)*

Middle. On 18 March 1950 the badly damaged No. 61057 is seen at Stratford awaiting a decision as to its fate. At the time of the accident it was less than 4 years old and became the shortest lived of the 410 strong B1 class. Although members of the class were still under construction at the time of the accident the missing number in the sequence caused by the demise of 61057 was never filled. *(Neville Stead Collection)*

Bottom. The following month it would be condemned with the locomotive and tender being cut up although the boiler was re-used. The severe damage to the cab is evident in this rear view. *(Neville Stead Collection)*

Ocean Terminal

Crowds gather on the rooftop viewing balcony of the Ocean Terminal to wave off departing passengers on the RMS *Queen Elizabeth* on another transatlantic run. A bevy of tugs stand ready to assist the great liner in easing her away from the dockside and proceeding down Southampton Water. Southampton is a deep water port with 4 tides daily and can accommodate the world's largest liners another of which, the RMS *Mauretania*, can be seen on the left. This Cunard liner stood in for the two Queens on the transatlantic run when they were undergoing maintenance but latterly operated cruises from New York. *(Donald Robertson)*

On July 31 the Prime Minister, Clement Attlee, opened a new passenger and cargo terminal at the UK's premier passenger port of Southampton. Named the "Ocean Terminal" the pre-cast concrete building stretching for nearly ¼ mile occupies almost the entire length of the east side of the Ocean Dock where the world's largest liners berth. A striking semi-circular tower rising to a height of 100 feet over four floors is situated at the seaward end of the main building which has two storeys. Passenger accommodation on the upper floor comprises two waiting halls adjacent to Customs Examination halls. These areas are linked to ships via electrically operated aluminium alloy telescopic gangways mounted in two turrets. There are six gangways arranged in three pairs carried on a housing mounted on the first floor landing platform. This housing can travel along the platform so that the gangways can be sited opposite the doors of the berthed liner and as each gangway is attached to a rotating turret it can be elevated or depressed to correspond with the height of the ship's doors.

The railway's island platform on the ground floor is more than 1000 feet long easily accommodating two full length boat trains. For those travelling by road there are road island platforms connected to the upper floor by escalator, stairway and lifts and when a passenger is ready to leave, his car is called by loud speaker from the adjoining car park. The terminal can process 2000 passengers an hour and those wishing to see loved ones depart are also well catered for by a viewing balcony that spans the entire length of the roof. Costing £750,000 to build it was designed by C.B. Dromgoole, architect to the BTC. The reception halls provided plush Vaumol hide seating, Vaumol being a trademark name for a method of colouring leather hide used by Connolly Leather until the mid-1980s. It was used chiefly for manufacturing leather upholstery for high class British automobiles of the era such as Rolls-Royce and Daimler. There was a refreshment buffet in each hall and there were telephone bays, complete with operators in attendance, where calls could be made to any part of the world. A flower shop, bookstall, writing room and an iced water fountain are all provided together with a Press Room for journalists. Eight exotic woods were used in the halls including Canadian wavy birch and bleached walnut. Electric automatic heating and air conditioning helped maintain an even temperature in the halls. As can be appreciated, no expense was spared to create an environment suitable to the class of passenger

OCEAN TERMINAL—FACTS AND FIGURES

Overall length .. .. 1,297 ft. 6 in.	Breadth of ground floor .. .. 111 ft. 6 in.
Height to ridge of roof 53 ft. 7 in.	Height to roof top of sightseers' enclosure 56 ft. 6 in.
Height of South Tower	79 ft. 6 in.
Height to Tower flagstaff top ..	102 ft. 0 in.
First Class Waiting Hall	221 ft. long by 102 ft. 6 in. wide by 22 ft. high
Cabin Class Waiting Hall	201 ft. long by 102 ft. 6 in. wide by 22 ft. high
First Class Customs Hall	465 ft. long by 94 ft. 6 in. wide by 32 ft. high
Cabin Class Customs Hall	344 ft. long by 94 ft. 6 in. wide by 32 ft. high
Total volume of building	6¼ million cubic feet
Steel used in construction	over 2,750 tons
Glass in windows	1½ acres
Barriers and guard railing	over 3,500 ft.
Artificial lighting	710 fluorescent lights; 12,500 ft. fluorescent tubing and 850 electric lamps

Broadcasting equipment includes 75 loud-speakers.
Structural steelwork disposed in transverse frames at 20 ft. 2 in. spacing.

The building is carried on 628 reinforced concrete piles, the majority of which are driven to 35 ft. below ground level and some down to 70 ft.

Train Island Platform. 1,010 ft. long, 30 ft. wide. External ferro-concrete canopy 1,048 ft. 6 in. long, 11 ft. wide, has 1,150 glass lenses of 9 in. diameter to admit daylight when train is standing at platform. Communication between inner and outer platforms is by way of 37 roller shutter doors 17 ft. 2 in. wide, operated manually or by mobile power unit.

Lifts. There are twenty-one connecting the ground and upper floors—eight from rail platform to each Customs Hall, two from each road island platform and one in South Tower.

Escalators. Four, one connecting rail platform to each Waiting Hall and one from each Customs Hall to road island platform. These four escalators are designed to deal with 4,000 persons per hour in either direction.

Baggage Conveyors. Two from each Customs Hall to ground floor cargo area, inclined at 30° and running at 70 ft. per minute.

Telescopic gangways can be swung out from balcony through 95° and elevated or depressed to an angle of 20° by means of hydraulic rams, an electric pump supplying the pressure. Telescoping is effected by electrically-driven endless chains. On landing outer end of gangway in ship's doorway all these motions are disengaged so that gangway may follow freely movements of ships due to tide, etc. Each pair of gangways with the shore housing weighs 17½ tons and contains 12,500 bolts and 100,000 rivets.

Heating. Two Waiting Halls have a combined heating and ventilating system, electrically operated. Heating load 750 kilowatts—equivalent to 375 ordinary 2-bar radiators.

BRITISH TRANSPORT COMMISSION
DOCKS & INLAND WATERWAYS

Hobbs the Printers of Southampton

(Image courtesy Mike Ashworth)

Amongst a forest of dockside cranes the elegant circular tower of the Ocean Terminal is evident in this view taken from the water.
(Donald Robertson)

likely to be using it and as the Docks & Marine Manager, one R P. Biddle, said in the brochure produced to accompany the opening –

"Ocean Terminal is acknowledged to be one of the world's best passenger reception stations and is a tribute to British design and craftsmanship. There is a streamlined efficiency about the operations which the ocean passenger of today rightly expects and which instantly creates a favourable impression."

The first ship to use the terminal was the *Queen Elizabeth* on the day after opening with the last liner to dock being its successor the *QE2* in December 1980. (Ed: The opening was covered by Pathe News and a short clip can be viewed on the internet).

Since completion in 1911 the Ocean Dock has served the great liners that crossed the Atlantic. The original passenger and cargo sheds were single storey buildings but in 1946 it was decided to replace two sheds on the eastern side of the dock with a modern terminal. This building did not finally see the light of day until the post war age of austerity rather than opening in an age of art Deco glamour to which it was more akin and thus it was already some years out of its time upon its completion. The RMS *Queen Mary* left Southampton for the last time in 1967 and the following year RMS *Queen Elizabeth* was retired, the age of the "Great Queens" had come to an end.

Opposite. The lavish interior of the terminal is revealed in these images taken from the official brochure. Externally the terminal was just as impressive with a Lord Nelson class locomotive arriving with a boat train illustrating the ease of transition from train to terminal. *(The following images courtesy Mike Ashworth)*

Below. As the bows of the majestic liner swing round to enable the great ship to head out into The Solent the impressive length of the Ocean Terminal is revealed. *(Donald Robertson)*

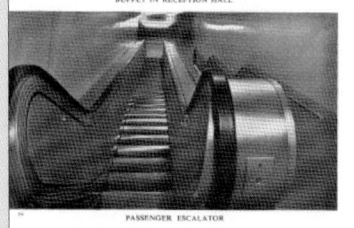

In 1952 British Transport Films (BTF) produced a 30 minute film showcasing the work of the Ocean Terminal and amongst the personalities featured are a tug skipper and his crew, a nursing stewardess on a ship bound for the Cape, an assistant in charge of baggage handling and freight, a taxi driver and a pilot taking a great liner down Southampton Water at night. The departure of the *Queen Elizabeth* and the arrival of the *Queen Mary*, one of whose passengers is Gracie Fields, are also covered and for the railway enthusiast there is the sight of Lord Nelson class 4-6-0 No. 30854 *Sir Martin Frobisher* leaving for the capital with a Pullman car boat train, S15 class No. 30498 hauling an outbound freight service, No. 34054 *Lord Beaverbrook* arriving with another Pullman boat train and several of the ubiquitous USA tanks shunting. BTF followed up this film in 1964 with another look at the port in a feature entitled "Southampton Docks" but this time concentrating upon the freight side of things, although steam was still in evidence with 9F No. 92155 seen hauling a van train out of the docks, although the boat train seen leaving the Ocean Terminal now has a Crompton diesel in charge.

(Ed: Sadly the days of the great transatlantic liners would not last long and by the 1960s inroads were already being made by airlines into this lucrative market. The Ocean Terminal closed in 1980 and although it was surveyed the following year by Television South, then due to be the new local ITV television franchise holder, with a view to conversion to studios this did not occur and the building was sadly demolished in 1983. Had it survived and been refurbished it would perhaps have made a more prestigious facility for the burgeoning cruise market of today in preference to the new Ocean Terminal building, constructed in 2009, which has little of the style of the original).

Timetable Expansion in size - and Price!

Following the edict that all timetables should conform to a new standard size from Nationalisation, the rules have been relaxed as from the summer timetable for 1950. Since 1948 timetables have been produced for all regions of BR in the standard format of that previously adopted by Bradshaw's Guide. With the introduction of the summer timetables for 1950 the LMR and the NER have increased the size of their timetables to 9in. x 6 in. The WR and SR continue to maintain the Bradshaw size whilst the ScR uses reprints from Murray's ABC Time Table. Notice that the opportunity was taken to hike prices for the NER timetable by 100% from sixpence to one shilling. The initial NER timetable in 1948 had cost just 3d – just goes to show rampant inflation is nothing new!

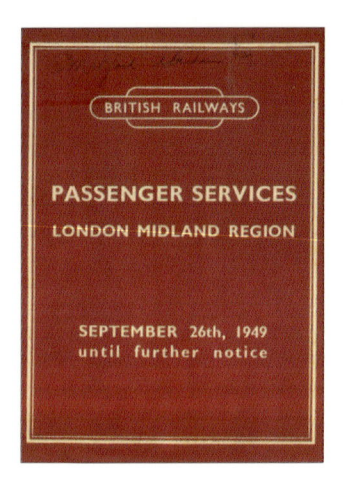

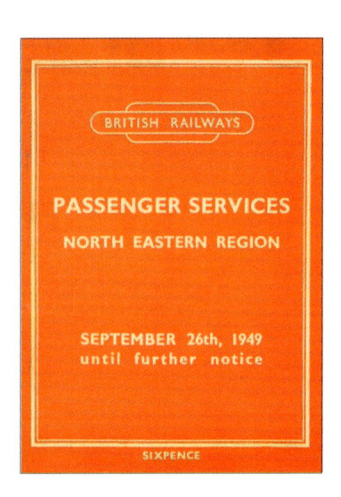

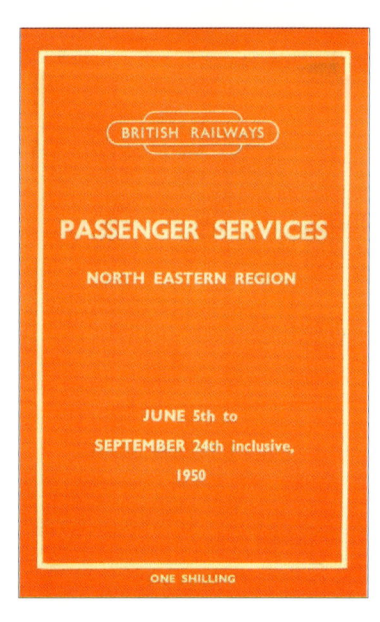

THE ScR TIMETABLE FOR 1950 REMAINED UNCHANGED IN SIZE AND IN PRICE USING REPRINTS FROM MURRAY'S ABC TIMETABLE

The Red Dragon

The "Red Dragon" named express was introduced by the WR on 5 June 1950, departing Carmarthen at 07:30 for Paddington, returning at 17:55. However in practice the main train started and terminated at Swansea with only a through portion working west of there to Carmarthen. Haulage was by Castles at first, then by Britannias, both types being supplied by Cardiff Canton depot. The name was withdrawn on 12 June 1965 but resurrected in various guises in later years.

The Inaugural up run saw No. 7018 *Drysllwyn Castle* haul the train from Carmarthen to Swansea where reversal took place with No. 5081 *Lockheed Hudson* handling the leg onwards to London. The name of the express was chosen to reflect the fact that the ancient battle standard of the Welsh was The Red Dragon (Y Ddraig Goch) consisting of a red dragon passant (i.e. standing with one foot raised), on a green and white background. As with any ancient symbol, the appearance of the dragon changed over the years, hence several different variations exist and in fact the headboards used by BR were to go through a number of changes. The first design was a BR Type 3 headboard, in black or red with polished aluminium lettering which was introduced in the summer of 1951. In 1956, a reversed style of painting was briefly used, with dark painted letters on a light background, still using the Type 3 design whilst in 1956 came a third style, perhaps the most well known, which remained in use until 1962. This comprised a curved rectangle, without cut outs to the upper corners, and was painted cream with brown lettering. In the upper centre, a disc protruding above the main headboard carried a moulded figure of a red dragon. A final design was used experimentally in late 1961 consisting of a rectangular fibreglass lightweight plate intended for diesel haulage.

Locomotive changeover at Cardiff General as an unidentified Castle class locomotive which had brought in the up "Red Dragon" from Swansea comes off the train to be replaced by Britannia pacific No. 70026 *Polar Star* for the onward journey to London. Both locomotives carry the headcode 720 representing this 7:30am service from Carmarthen to Paddington. This image dates from 1 June 1953. *(R. C. Riley)*

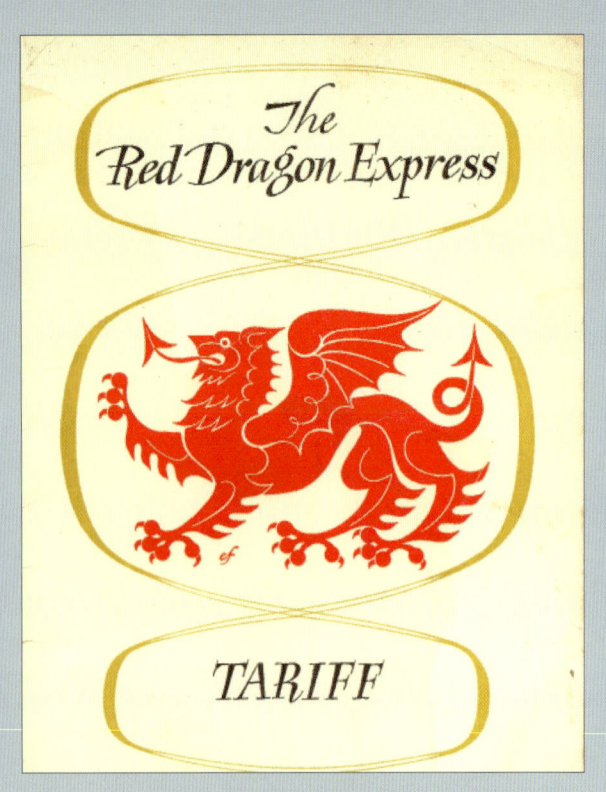

BR booklets produced in the 1950s.
(Courtesy Mike Ashworth)

In the other direction the route number for the down "Red Dragon" was 173 as worn here by Castle class No. 5089 *Westminster Abbey* seen alongside Britannia No. 70018 *Flying Dutchman* at Paddington on 4 August 1952. *(R. C. Riley)*

'Kerosene Castle' - The UK's Prototype Gas Turbine Locomotive

In 1946 the GWR had ordered a gas turbine locomotive from the Swiss firm of Brown – Boveri to be numbered 18000. The electrical engineering company of Brown, Boveri & Cie. was founded in 1891 in Zurich by the gloriously named Charles Eugene Lancelot Brown together with Walter Boveri who worked for Maschinenfabrik Oerlikon another Swiss engineering company based in Zurich's Oerlikon district who were known for their early development of electric locomotives. After extensive trials in Switzerland No. 18000 (which was to be designated GT1) arrived in the UK via the Harwich ferry on February 3 1950 and after being towed to Swindon underwent trials before entering service at Old Oak Common in September. Slightly before the order had been placed for the Swiss locomotive a similar gas turbine locomotive had been ordered by the GWR from the UK's Metropolitan Vickers Company, to be numbered 18100. However delivery of this locomotive was delayed with No. 18100 (designated GT2) not arriving until December 1951.

A series of trials was run but neither locomotive was deemed to be a great success for as with all gas turbine units that operate at or close to sea level, their high fuel consumption under part load conditions made them very expensive to operate with, for example, the Metrovick's fuel consumption, using expensive kerosene aviation fuel, averaging 2.97 gallons/mile (almost roughly three times that of an equivalent diesel electric engine). Additionally they were subject to frequent breakdowns and spent much of their short working lives inside Swindon Works. 18100 only worked as a turbine for a very short time and suffered particularly from stress cracks to its bogies. It was withdrawn from service in December 1952 and returned to the makers at Dukinfield in an attempt to convert it to burn the less expensive heavy fuel oil, the same type as powered 18000. This attempt at conversion lasted only from 1953 until 1957 when the idea was abandoned. Both machines were noisy and unpopular with the crews such that 18100 was withdrawn in 1958 after only six years in service with No. 18000 following in 1960. This unpopularity with staff was reflected in the nickname "Kerosene Castle" bestowed upon No. 18000. Ultimately the locomotive was returned to Switzerland where, after rebuilding, it was used as a mobile rail adhesion testing unit for the 'Union Internationale des Chemins de Fer' in Vienna acquiring the rather more endearing name of 'Elisabetta' which the test team there bestowed on it in recognition of its origins . Returning to

No. 18000 is seen at Taunton on 20 September 1950. *(Neville Stead Collection)*

Right. Swindon Works hosts a visit from the first of the Brown-Boveri gas turbines No. 18000. *(Henry Cartwright)*

the UK in 1995 it was initially displayed at The Railway Age, Crewe but is now located at Didcot where in 2022 it was announced that work would begin on a two year conservation project to repair corrosion on the bodywork and to return the locomotive to the attractive black and silver livery it carried when first in service.

After No. 18100's withdrawal from service early in 1958 it was stored at Swindon Works before being returned once more to Metropolitan Vickers for conversion as a prototype 25kv AC electric locomotive. It emerged numbered E1000, later changed to E2001, and was given the TOPS classification of class 80. It was scrapped in 1972. This was not quite the end of the gas turbine story however for English Electric built a demonstrator in 1961 designated GT3. This was essentially a standard oil-fired gas turbine mounted on a standard steam locomotive chassis with a tender filled with kerosene which, although it sounds rather crude, did enable the design to avoid much of the unreliability which had plagued the earlier complex experimental Nos. 18000 and 18100. However, it failed to be competitive against conventional diesel electric traction and after completion of several test runs over Shap GT3 was returned to English Electric at the end of 1962 and stored. The locomotive was then partially dismantled and its turbine and heat exchanger equipment removed being finally scrapped by Ward's of Salford in February 1966.

Seen outside a snowy Swindon Works on 30 March 1952 the second example of the gas turbine locomotives No. 18100 was paying one of its frequent visits here for attention. *(Leslie Freeman)*

On test with a solitary coach in the West Country GT2 No. 18100 climbs Dainton Bank on 9 February 1952. *(Peter Gray)*

Looking smart in its black livery with silver stripe and cast metal numbers GT2 No. 18100 makes ready to take out a service from Paddington. *(Milepost)*

In its reincarnated state the former GT2 is now an electric locomotive numbered E2001 and is seen here at Rugby on 29 March 1965. *(Railway Image Collection)*

Under the shear legs at New England shed in the summer of 1961 is GT3 the last of the trio of gas turbine locomotives. The design conformed very much to a streamlined steam locomotive outline with a separate tender. *(Barry Richardson)*

Around the Regions in 1950

The "Devon Belle" all Pullman train makes its way gingerly across the Taw viaduct between Barnstaple Town and Junction stations on 26 August 1950 with Bulleid Pacific No. 34034 *Honiton* in charge. *(Neville Stead)*

On 13 July 1950 Class D30/2 No. 62422 *Caleb Balderstone* is captured at Newcastle's Central station passing under the station's impressive signal gantry. One of the famous "Scott" class it was named after a character from the novel "The Bride of Lammermoor". *(Neville Stead)*

RAILWAY TIMES 1950

Sentinel No. 47190 is on shed at its home depot of Bristol Barrow Road on 24 September 1950. In August 1952 it would be moved to the S&D for use at Radstock to join classmate No. 47191. *(Neville Stead Collection)*

No. 4 one of BR's Grimsby-Immingham 500v DC electric tramcars is seen on 18 March 1950. Supplied by Brush Engineering for the GCR these cars could take 40 passengers seated with another 30 or so standing the clientele being mainly dock workers travelling from Grimsby to the docks at Immingham. The last section of the tramway to remain open closed on 1 July 1961 following road improvements. *(Neville Stead Collection)*

WR 655 class pannier tank No. 1782 engages in some shunting at Truro on 9 April 1950. This 0-6-0 would be withdrawn from here in November that year. *(Peter Gray collection)*

RAILWAY TIMES 1950

In 1950 Fraserburgh shed plays host to D41 Class No. 62230 dating from 1893. This 4-4-0 veteran would end its days at Keith shed in 1952 after 58 years service. *(Neville Stead Collection)*

Class O2 No. 14 "Fishbourne" waits with its two coach branch train at Bembridge ready to return to the junction at Brading in 1950. *(Lens of Sutton, Dennis Culum Collection)*

Last Survivors
Around the regions several classes had their last examples withdrawn during 1950

WR
No. 1331 was the last surviving locomotive of the Cardigan & Whitland Rly. This 0-6-0ST built by Fox Walker in 1877 was acquired by the G.W.R. in 1886. It had a service life of 73 years latterly being allocated to Oswestry shed. It was originally numbered 1387 by the GWR and transferred to departmental stock in 1902. Following overhaul in 1926 it was renumbered 1331 and added to running stock in November 1927.

Above, Ex Whitland & Cardigan Rly. 0-6-0ST No.1331 now in the GWR fold inside Oswestry MPD seen in 1946 *(Arthur Mace)*

By 1950 No. 1331 had reached Swindon Works from where it would be scrapped later that year. *(Neville Stead Collection)*

RAILWAY TIMES 1950

SR
The former K&ESR 0-8-0T "Hecate" was withdrawn from Nine Elms in March.

Ordered from Hawthorn Leslie in 1905 at a time when the K&ESR planned to extend to Maidstone, it proved rather too heavy for the line's track and was exchanged in September 1932 for an old SR "saddleback" 0-6-0T. Numbered 949 by the SR (who also replaced the boiler in 1939) it then continued to find work at Nine Elms until 1950 when it sustained damage after being involved in an accident and scrapped.

Hecate the former K&ESR 0-8-0T is seen here at Nine Elms in 1948. *(Alec Ford)*

ER/NER
The last two Great Central Class D9 Nos. 62305, 62307 have been withdrawn.

This class of 40 4-4-0s was designed by Robinson and constructed between 1901-4 being later rebuilt with larger boilers. 26 were taken into BR stock in 1948 mainly concentrated in Cheshire. No. 62307 *Queen Mary*, based at Stockport Edgeley, was one of the last pair to survive into 1950 and was also one of only four members of the class to be named.

Class D9 No. 62307 is seen at Trafford Park in February 1950. *(Flint and Herbert)*

RAILWAY TIMES 1950

No. 62305 was photographed working an unidentified service at an unrecorded location. *(Peter Pescod)*

MR
The last ex MR 2-4-0 No. 20216 has been withdrawn

The Johnson 1400 Class 2-4-0s were considered by many to be one of the most elegant of the wide variety of Midland Railway locomotives of this wheel arrangement. Constructed between 1879 and 1881, thirty were produced at Derby whilst a further thirty were manufactured by Neilsons. No. 20216 retained its LMS number until withdrawal.

The elegant lines of this LMS ex MR 2-4-0 are evident as No.20216 waits at Cheltenham Lansdown station in 1949. *(Milepost)*

The last 2-4-0 and the oldest tender passenger locomotive on the LMR has met its end

No. 20155 was withdrawn in October, without having received its BR No. of 58020, bringing down the curtain on Johnson's 1070 class this example having been constructed for the MR at Derby in 1876. It was destined to be the last 2-4-0 and the oldest tender passenger locomotive remaining on the LMR.

Above. Another view of this 1P class No.20216 in the goods yard at Cheltenham Lansdown. *(Milepost)*

Below. This Johnson 2-4-0, which would be withdrawn in October 1950, is seen here at Nottingham on 24 June 1949 destined to be the last of its class to remain in service. *(J. S. Cockshott)*

10201 - First SR Mainline Diesel Completed at Ashford

The first of a trio of prototype diesel locomotives built by BR at Ashford Works emerged in 1950 with the second following in 1951 and a third in 1954. Readily identifiable as having been designed by Bulleid they had been authorised back in 1947 by the Southern Railway. The engine and transmission were supplied by English Electric but the style of the body shell was pure Bulleid closely resembling his electric locomotives and being markedly different from the typical English Electric style which had been heavily influenced by contemporary American design. On December 5 ten empty corridor coaches were hauled by 10201 on a trial return run from Ashford via Dover, Minster and Canterbury. As early as the following month the locomotive journeyed to the LMR where it was allocated to the Derby – St. Pancras service. No. 10201 would later be returned to Ashford for final painting and fitting out before being exhibited at the 1951 Festival of Britain on the South Bank. It would enter full time revenue earning service in September 1951. Following delivery of the second locomotive No. 10202 was trialled on Waterloo - Salisbury and Exeter services. Once both locomotives became fully operational they were deployed on Exeter and Bournemouth diagrams working alternately in order that one was always available as a standby in case of breakdowns.

Illustrating their use on the West of England mainline first of the trio No. 10201 is seen passing a rather rain soaked Vauxhall with a service for Plymouth. *(Ken Coursey)*

They were also put to use on the Bournemouth route and here No. 10201 is captured near Brockenhurst on 24 May 1953. *(R. C. Riley)*

This impressive front end view of No. 10201 shows off the elegant styling of the design and was taken at Clapham Junction on 3 March 1952. *(Roy Vincent)*

Back on the West of England mainline but this time with classmate No. 10202 seen here departing from Salisbury with an Exeter service. *(Flint and Herbert)*

No. 10201 is here entrusted with the prestigious 1:10pm "Golden Arrow" Pullman service from Victoria to Dover seen here passing Herne Hill signalbox in 1954. (5002A)

Lauder Branch Re-opens for Freight

The neat well kept terminus at Lauder seen on 10 July 1952 during the years after re-opening to freight following temporary closure by the East Coast floods of 1948. *(Neville Stead Collection)*

November 20 saw the re-opening of the freight only Lauder branch from the junction on the Waverley route at Fountainhall. Closed as a result of the August 1948 floods having damaged the bridge over the Gala Water near Fountainhall, repairs have taken some time owing to priority being given to restoring the damage on the ECML, details of which were covered in Railway Times Nos. 1 and 2. Problems in obtaining regular supplies of materials have delayed reconstruction of the bridge until now. Opened in 1901 as a light railway, passenger services were an early casualty having been withdrawn in 1932. (Ed: Freight services would continue until 15 November 1958 when the final enthusiasts' train was hauled by Standard Mogul No. 78049).

Right. Oxton was one of the intermediate stations on the branch and is seen here in July 1952. *(Neville Stead)*

Ready to haul the final passenger working to Lauder from Fountainhall Junction on the Waverley route, is Standard Class No. 78049 suitably adorned with headboard on 15 November 1958. *(W. A. C. Smith)*

The Provost of Lauder is about to place a wreath on the front of No. 78049 operating the Branch Line Special of 15 November 1958 after arrival at Lauder. *(W. A. C. Smith)*

The Welsh Dragon

In addition to the inaugural run of the Castle hauled "Red Dragon" express another mythical Welsh beast reared its head this year, although this time in the form of a pull-push set named holiday train operated by a tank locomotive!

On July 3 the MR introduced a new holiday service running between Llandudno and Rhyl in North Wales. Although not shown in public timetables the service ran on Mondays-Fridays only, Saturday being the traditional holiday changeover day and of course a day of peak occupation on the mainline, and comprised seven trips daily in each direction which, by 1953, had increased to eight and additionally six departures daily were offered on Sundays. The last departure from Llandudno was not until 10:45 pm thus giving tourists the chance to avail themselves of the shows on offer at the resort. A pull-push set made up of two or three coaches conveyed the holidaymakers for the princely sum of 2/6d return (12½p). The locomotive, usually one of Rhyl depot's Ivatt 2-6-2 tanks, carried a headboard consisting of a red dragon upon a green shield above the legend "The Welsh Dragon" finished in yellow lettering on a red ground. When the locomotive was pushing the headboard was mounted on the front of the leading carriage. The name was also displayed on carriage boards carried on the first and last coaches of the train. It was reported that the trains were well patronised as the fare compared favourably to that charged by local bus services. Surely unique amongst named trains in two respects – using a pull-push set and haulage by a tank locomotive, "The Welsh Dragon" is certainly something of a "one off". Finishing for the season on 15 September it is hoped that a similar service will run in future years. (Ed: The service was resurrected in subsequent years although by the summer of 1957 Derby Lightweight DMUs had taken over before a brief return to steam in 1962).

This view of the rear coach and propelling locomotive was taken near Colwyn Bay. The locomotive usually propelled from the Llandudno end of the journey with the 17½ miles to Colwyn Bay scheduled for between 31 and 39 minutes. *(Roy Vincent)*

RAILWAY TIMES 1950

THE
WELSH DRAGON
DIESEL TRAIN EXPRESS SERVICE

MONDAYS to FRIDAYS, 12th JUNE to 8th SEPTEMBER
SUNDAYS, 25th JUNE to 10th SEPTEMBER

	MONDAYS TO FRIDAYS								SUNDAYS						
	a.m.	noon	p.m.	p.m.	p.m.	p.m.	p.m.	p.m.	p.m.	p.m.	p.m.	p.m.	p.m.	p.m.	p.m.
Llandudno	10 30	12 00	1 30	4 00	6 00	8 05	9 03	10 45	1 28	3 00	4 45	6 15	7 45	10 00	
Deganwy	10 35	—	—	—	—	—	—	10 50	—	—	4 50	—	7 50	—	
Llandudno Jct.	10 40	12 08	1 38	4 08	6 08	8 13	9 08	10 54	1 38	3 10	4 56	6 35	7 55	10 08	
Colwyn Bay	10 46	12 15	1 45	4 15	6 15	8 24	9 15	11 01	1 37	3 17	5 02	6 32	8 02	10 15	
Abergele	10 55	12 25	1 55	4 25	6 25	8 33	9 24	11 10	1 47	3 27	5 12	6 42	8 12	10 25	
Rhyl	11 03	12 32	3 03	2 32	4 32	8 41	9 32	11 18	1 54	3 34	5 19	6 49	8 19	10 32	

	a.m.	a.m.	p.m.	p.m.	p.m.	p.m.	p.m.	p.m.	p.m.	p.m.	p.m.	p.m.	p.m.	p.m.
Rhyl	9 45	11 20	1 45	3 10	5 10	6 55	8 50	9 50	2 10	3 45	4A 45	5 35	7 00	8 30
Abergele	9 51	11 26	1 51	3 16	5 16	7 01	8 56	9 56	2 16	3 51	4A 51	5 41	7 06	8 36
Colwyn Bay	10 01	11 36	2 01	3 26	5 26	7 11	9 06	10 06	2 26	4 01	5A 01	5 51	7 16	8 46
Llandudno Jct.	10 08	11 43	2 08	3 33	5 43	7 18	9 12	10 12	2 34	4 07	5A 08	5 57	7 23	8 52
Deganwy	—	—	2 13	3 38	—	—	9 18	10 18	2 38	4 12	5A 14	6 02	7 27	8 57
Llandudno	10 20	11 52	2 19	3 44	5 51	7 27	9 24	10 22	2 44	4 18	5A 20	6 08	7 33	9 03

This is not a complete list of the services between Llandudno and Rhyl. Details of these and other trains may be obtained at the station.

NOTES. A—Commences 18th June.

41285 is seen bowling along in fine style near Colwyn Bay resplendent with headboard. Although some turnround times were tight, the service seems to have enjoyed a good punctuality record in its first year of operation. *(Roy Vincent)*

Festival Preparation
BR Announces Plans for Festival of Britain Services in 1951

The Festival of Britain was a temporary exhibition running for 5 months in London from May to September 1951. It generated a profit, was very successful in showcasing Britain and proved extremely popular with the public with some 8.4m visitors. **(Ed: I remember being taken to see it as a very young child and being particularly fascinated by the Skylon, the futuristic metal sculpture that came to symbolize the exhibition site on the South Bank of the Thames).**

In late 1950 BR got ahead of the game and announced its plans with regard to trains serving the capital aimed specifically at tourists and UK residents visiting the exhibition. One can do no better than quote from the above brochure which set the scene thus –

"The theme of the exhibition is Britain her past, present and future. As a major partner in the field of transport British Railways have an important role in this historic presentation. This brochure outlines something of our plans to speed both the arriving guest and those of our own countrymen who will be eager to visit the many exhibitions and festivals which will be staged throughout Great Britain." London, centrepiece of the Festival will have the Exhibition at the South Bank as its main attraction but there will also be pleasure gardens in Battersea Park with open air cafes and restaurants together with music and entertainments. Ravaged by bombs in the recent war Poplar, where a new town the Lansbury Estate is being built, will act as an exhibition of architecture. South Kensington will hold an exhibition of science and there will be a book exhibition at the Victoria & Albert Museum from Caxton's time to the present day which will also house a display featuring the 1851 Great Exhibition the centenary of which is being commemorated by the 1951 festival. All in all it was promised that "the capital will be gay with decorations, fountains, floodlighting and fireworks."

In addition there will be travelling exhibitions by land visiting Manchester, Birmingham, Leeds and Nottingham, by sea visiting Newcastle upon Tyne, Hull, Southampton, Plymouth, Bristol, Birkenhead, Dundee, Glasgow and Cardiff. There will be festivals of the arts in a number of towns including Aldeburgh, Bath, Cambridge, Canterbury, Cheltenham, Oxford and Stratford - upon - Avon. Scotland is also scheduled to have exhibitions in Glasgow's Kelvin Hall and in Edinburgh whilst Wales will have arts festivals in Llangollen, St. David's, Llanwrst and Swansea and Northern Ireland will host

The bustling crowds which flocked to the festival site are captured here beneath the supporting struts of the Dome of Discovery, the largest dome in the world at that time with a diameter of 365ft. The Transport Pavilion which consisted of four sections covering sea, road, rail and air, is in the background. *(Roy Vincent)*

exhibitions and arts festivals in Belfast.

Tapping into the anticipated increase in rail travel during this period BR took the opportunity to showcase the new standard passenger coaches and restaurant cars often hauled by the new Standard class of locomotives. New Pullman cars will also be introduced on the "Golden Arrow" service, a vital gateway for overseas visitors, during the festival season. In addition to the existing "Royal Scot" and "Norfolkman" expresses, which will also operate with the new stock, five new named trains carrying names relevant to the route they operate and featuring these innovations will be introduced.

They are -

The leaflet produced for the "Red Rose" for example went into some detail regarding the new stock – *"The coaches, 63ft. 6" in length, have all-steel bodies and are equipped with automatic couplers and Pullman type gangways. Moquette upholstery is used throughout and decorative woodwork. Chairs are moveable in the restaurant cars, the seating arrangement being two persons to each table on one side of the gangway and four to each table on the other. The kitchen cars are the largest ever put into service by British Railways, with alternative cooking by anthracite or electricity".*

THE RED ROSE	London Euston to Glasgow Central
THE HEART OF MIDLOTHIAN	London King's Cross to York and Edinburgh Waverley
THE MERCHANT VENTURER	London Paddington to Bath and Bristol
THE WILLIAM SHAKESPEARE	London Paddington to Birmingham, Wolverhampton and Stratford-upon-Avon
THE ROYAL WESSEX	London Waterloo to Winchester, Weymouth, Swanage and Bournemouth

In addition to these new named services the "Thanet Belle" will be renamed the "Kentish Belle" and will additionally serve Canterbury. (See feature in RT1 p26). The SR will also benefit from a new nightly midnight service from Victoria to Brighton with connections for Eastbourne, Hastings and Worthing, and on certain days of the week a late service will operate from Charing Cross to Ashford, Folkestone, Dover and Deal and to Canterbury, Faversham and the Kent coast whilst on Wednesdays and Saturdays Sevenoaks and Tunbridge Wells will be catered for. These additional services would no doubt be a great boon to the weary festival goer in their trek home after a very full day at the exhibition.

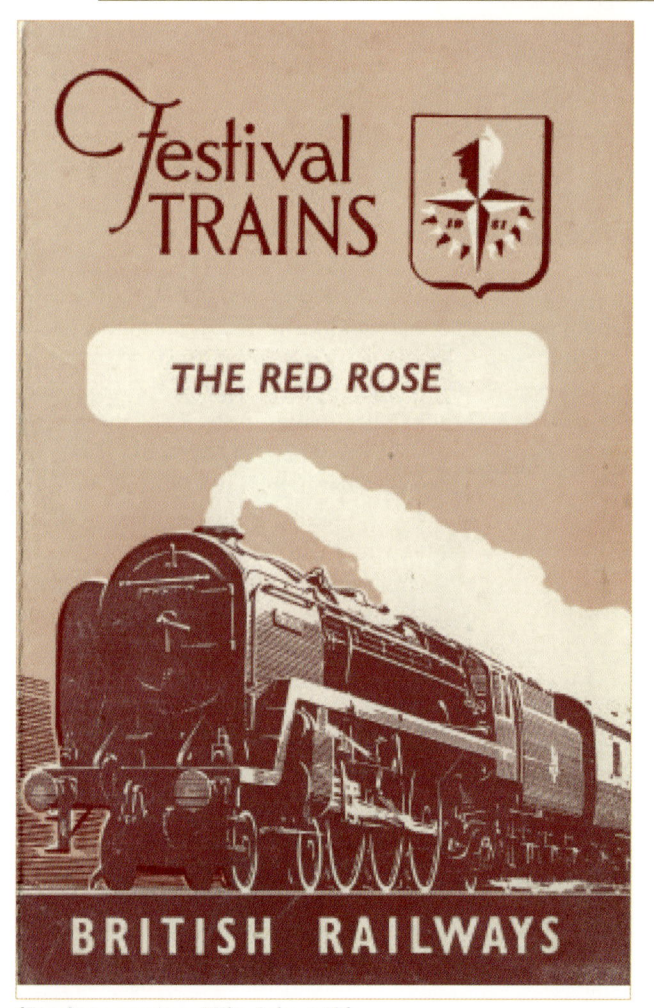

(Brochure courtesy Mike Ashworth)

Bottom right. The other worldly quality of the Skylon is particularly effectively captured in this nocturnal shot of the South Bank site. *(Roy Vincent)*

RAILWAY TIMES 1950

One of the new services introduced to mark the Festival of Britain was the "Heart of Midlothian" express seen here passing Doncaster on the up service with Peppercorn A2 pacific No. 60538 *Velocity* at its head. *(Roy Vincent)*

The new service also carried its designation on the end plate of the rear carriage seen here passing Arksey just to the north of Doncaster on its down journey to Edinburgh during the height of the festival season in June 1951. *(Roy Vincent)*

RAILWAY TIMES 1950

Captured on 20 August 1950 this tramcar LT2093 was working the "Emergency shuttle service" introduced to mitigate the effects of the road works under way in preparation for the festival. In spite of the abandonment of trams in the capital being well in hand by this time they were to play an important role in transporting passengers to and from the festival in the following year with the area around Westminster and the Embankment being strongholds of the remaining tram routes well into 1951. The last tram was to run in 1952. *(John Meredith 137-1)*

In preparation for the festival this image, dating from 17 February 1950, shows what was claimed to be the last piece of new conduit track to be laid down anywhere in the world. This was a siding installed in Addington Street beside County Hall over several months in early 1950. The conduit system allowed current to be collected at ground level by a "plough" rather than from overhead wires. *(John Meredith 99-12)*

(Ed: Details of the many railway exhibits to be found in the Transport Pavilion will be included in the next issue of Railway Times).

Brief Encounters

SPRECHEN SIE DEUTSCH AT PARKESTON QUAY ?
The Eastern Region of BR has introduced multi-lingual train announcements at Parkeston Quay and on its "Hook Continental" boat train. This has been done to speed up Customs and Immigration formalities and to assist visitors from overseas. Using magnetic tape a selected range of standard announcements has been recorded and there is also the facility for making special "one off" announcements as required. English, Dutch and German are used for services to and from The Hook of Holland whilst English and Danish feature for the Scandinavian services. (**Ed:** The following selections are probably announcements you would be unlikely to hear however) –

> **Die falsche Art von Schnee**
> **Sehen, sagen , sortiert**
> **Bladeren aan de lijn**
> **Vær opmærksom på, at der ikke er nogen taktil belægning på denne platform**

Or in other words –

> **The wrong kind of snow ****
> **See it, say it, sorted**
> **Leaves on the line**
> **Please be aware there is no tactile paving on this platform**

** To be fair to BR this quote was in fact originated by a BBC reporter on the "Today" programme rather than the BR interviewee whose words were actually **"We are having particular problems with the type of snow, which is rare in the UK"**, the snow being of a powdery nature and far colder than usual. Christian Wolmar in his book **"British Rail – the Making and Breaking of our Trains"** rightly points this out as being one of those myths such as "curly sandwiches" with which BR was saddled.

"The Scandinavian" boat train express ready to depart from Parkeston Quay for Liverpool Street with Class B12/3 No. 8559 in LNER colours at the head. This view dates from the period between renumbering from 7383 to 8559 in 1946 and the later renumbering to 68859 in the BR series in 1949. *(Dr. Ian C. Allen)*

"The Hook Continental" makes a stirring sight as it leaves Liverpool Street with the boat train for Parkeston Quay with No. 70001 *Lord Hurcomb* in charge on an unrecorded date in 1951. Passengers will no doubt be treated to on board announcements courtesy of the new equipment installed but hopefully with less frequency than those annoying announcements which we have to endure these days. *(Roy Vincent)*

RAILWAY TIMES 1950

WHAT'S IN A NAME

From having one of the shortest names carried by locomotives of the "West Country" class, a distinction it shared with No. 34006 *Bude*, No. 34092 *Wells* has been adorned with new nameplates *City of Wells* as from March. Since it was originally named on 25 November 1949 at Priory Road station in the city by the Mayor Alderman Kippax, with the Bishop of Bath & Wells in attendance, residents of the cathedral city, the smallest free standing city in the UK, have campaigned for a name change to incorporate the city's status. BR have finally relented and in March of this year the new plates were mounted. (Ed: Members of the Wells Railway Fraternity travelled to Bury on 25 November 2015 for a rededication ceremony for *City of Wells* after 25 years out of service undergoing overhaul, 66 years to the day after its first naming back in the Somerset city).

Pride restored – well at least in giving Wells its city status – although this image of a severely work stained external appearance does not say much for presentation! (Barry Richardson)

Passing Chapelton, the first station south of Barnstaple, in 1956, hopefully relatively quietly, No. 34014 Budleigh Salterton is in charge of the 2-20 pm train from Ilfracombe to Waterloo (Lens of Sutton, Dennis Cullum Collection)

PIPE DOWN

The Doppler effect of a steam whistle from a locomotive passing at speed is surely one of the great sounds to be experienced on a steam railway however on the evening of June 29 it became something of nuisance for travellers on an Exeter to Ilfracombe service. The culprit was No. 34014 *Budleigh Salterton* whose crew vainly tried to silence the whistle which refused to shut off on departure from St. David's station on the 54 mile journey to the North Devon resort. It whistled continuously for the next 3 hours and must have driven the crew mad but after picking up a fitter at Barnstaple Junction, silence reigned supreme or at least it did for a while because on the return journey the whistle decided to operate again but this time of its own accord. Reaching Barnstaple at about 9pm the fitter again "had a go" (clouting it with a hammer I suspect) and succeeded in finally closing the valve. Although the pacific had received a light casual overhaul at Eastleigh Works only the previous month, 5 May – 26 May 1950, it was subsequently revealed that the cause had been a bent whistle valve spindle.

OUT OF SIGHT OUT OF MIND

As detailed in Railway Times No.2, Leader's trials in 1949 and 1950 had not done enough to convince Robert Riddles that the development should proceed following the publication of a critical report that was issued in December 1950 after the completion of dynamometer trials. The decision to scrap the whole project would be taken in March 1951 but whilst the trials were ongoing with 36001 the order came from on high to cease any further work on the other locomotives. Thus the partially completed Nos. 36002 and 36003 were shunted off to less visible surroundings than residence outside Brighton Works, finding a home initially at Bognor Regis where they were reported as being on shed in May. Towards the end of the year they would both would be taken to New Cross Gate in London for storage and by June 1951 36003 was towed back to Brighton this time for scrapping.

The partially completed No. 36003 pokes its nose out of Bognor Regis shed whilst classmate No. 36002 skulks inside In May 1950. *(R. C. Riley)*

Far more visible to the travelling public was this parking spot outside Brighton Works where No. 36003 was kept for some time whilst its future was being debated. *(Neville Stead Collection)*

RAILWAY TIMES 1950

Moved from Bognor Regis to London's New Cross Gate shed No. 36002 was photographed on 23 June 1951 prior to being towed to Brighton for scrapping. *(Neville Stead)*

Perhaps appropriately on April Fool's day 1950 the only example to have run tests was to be found parked up at Brighton Works in rather woebegone condition after having spent several days "dead" on Brighton shed. Pending further modifications undertaken at the Works it was taken to Eastleigh for weighing and for the fitting of apparatus for use in subsequent dynamometer car tests. *(Neville Stead)*

RAILWAY TIMES 1950

ON THE ROCKS

A rather amusing leaflet issued on behalf of BR to assist American visitors to the UK in the post-war years offered suggestions regarding some of the available travel facilities on offer covering not only train travel but also motor coach, steamer and ferry services, tickets, fares and railway hotels. It also helpfully explained some of those "quaint little old British customs" that our American cousins might come across in their travels. The initial edition came out in 1948 but the one illustrated here dates from 1950. Attention is drawn to the usefulness of obtaining "Mileage Coupons" back home as they allowed an average 32% saving on fares in the UK. Other fare deals included London Transport "Go As You Please" tickets that at the time were restricted to overseas visitors. The intricacies of pounds, shillings and pence were explained as were such British foibles as the habit of serving drinks without ice which were covered thus - "If your drink is not full of tinkling ice cubes as you're accustomed to having it at home, remember, some of the British like it that way!". The British taste for "warm beer" was wisely not mentioned and of course one must always remember that refrigeration was not at that time widespread in the British Isles. Some of the cartoons employed the use of "Chad" the chalk drawn character used to emphasise wartime and austerity shortages with the opening slogan "Wot no...". The BR 'totem' logo makes several appearances, sometimes using the correct Gill Sans typeface although at other times with an Americanised slant on the lettering but I suppose at least BR was not referred to as "British Railroads"!

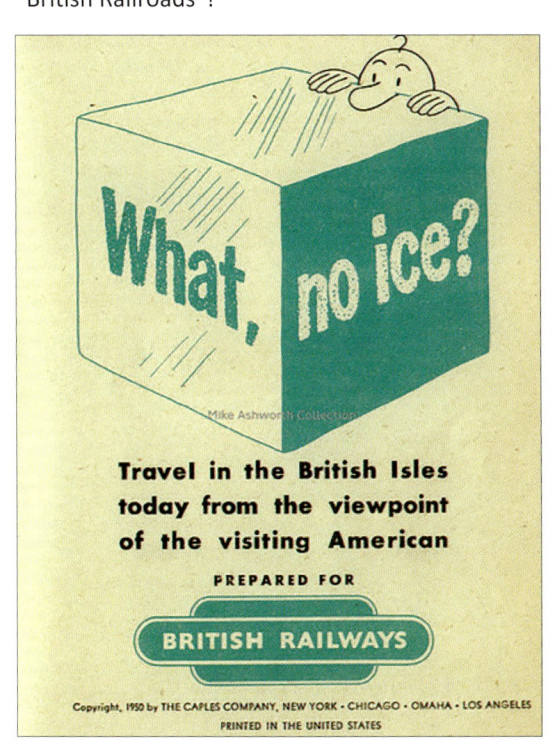

The 30.5% devaluation of the pound in 1949, which meant the previous exchange rate of $4.04 tumbled to $2.80, also gave rise to savings such as that mentioned here where a reserved seat for a 400 mile journey costs only 14 cents. *(Images courtesy Mike Ashworth)*

Also Making the News

The BR magazine has a new format as from January with each of the six regions having its own issue, although the first 4 pages of all editions will be devoted to a broader national approach encouraging staff to think in terms of BR as a whole. The introductory pages will explain what the thinking of the Railway Executive is on various matters.

To replace older locomotives on branchline construction work the first batch of 20 0-6-0PTs has recently started at Swindon. They will be known as the 1600 class.

Below. Delivered in June 1950 No. 1623 was one of the class fitted with a spark arrestor chimney for working the timber yards in Gloucester docks. It is seen here on Gloucester's Horton Road shed on 4 July 1954. It was to spend thirteen years based here from new until its final move to Llanelli in June 1963. *(Peter Gray)*

RAILWAY TIMES 1950

Left. Newly constructed 1600 Class No. 1653 seen at Swindon in pristine condition in December 1954. Delivery of these 0-6-0PTs had begun in October 1949 and continued until May 1955 by which time 70 had been delivered. *(John Bell)*

A plaque was erected on the Britannia Bridge linking Anglesey with the mainland on 21 October commemorating 100 years of service since its completion in 1850.

Liverpool Street has had the first pneumatic buffer stop installed.

One of the unpopular Bulleid tavern restaurant cars, No. 7836, has been modified to provide full size side windows in place of the former small high level windows. Seats have been turned to conform to the normal transverse arrangement rather than facing inwards.

Following experience gained with the operation of the Bulleid Double Deck EMUs it has been decided not to proceed with any further examples. BR has stated that longer trains and longer platforms are the preferred solutions as the trials have revealed that longer station dwell times have more than offset any advantage from the extra seating capacity. Adverse public reaction to uncomfortable seats, cramped space and ventilation problems, as mentioned in a previous "Railway Times" article, have also led to this decision as well as considerations of loading gauge restrictions involved in extending their use to other routes.

Stranraer – Portpatrick passenger service was withdrawn from February 6 largely due to the fact that the competing bus service was more frequent and faster than the limited service offered by rail.

The remains of Portpatrick station seen here in May 1965 some 15 years after passenger services were withdrawn. *(W. A. C. Smith)*

RAILWAY TIMES 1950

BR has issued a new comprehensive list of motive power depot codes which in summary are –

1A – 28B	LMR
30A – 40F	ER
50A – 54D	NER
60A – 68D	ScR
70A – 75G	SR
81A – 89C	WR

Right. 86F was the shedcode for Tondu in South Wales until 1960 when it changed to 88H, 86F then became Aberbeeg until that depot closed in 1964.

A new freight and passenger headcode classification was introduced from June 5 providing for 10 headlamp codes for general use on BR. However, it does not apply on the SR, where there are more than 30 different route combinations in use, nor on LTE lines.

Three new named expresses have been introduced on the Eastern Region -

The Tynesider	London King's Cross – Newcastle (Sleeper service)
The Easterling	London Liverpool Street – Lowestoft, Yarmouth South Town
The Broadsman	London Liverpool Street – Cromer, Sheringham

No. 61049 of Class B1 handles the up 'Broadsman' passing under the wires beside the Ilford flyover on 21 April 1951. This service from London's Liverpool Street ran to Norwich, Cromer and Sheringham until 1962. The shields on the headboard depict yachts and windmills on a royal blue background. Later in 1951 Britannia Pacifics were employed on the service cutting some 26 minutes from the journey time. *(Roy Vincent)*

RAILWAY TIMES 1950

Top left. This view is shortly after Britannias had taken over operation of "The Broadsman". Delivered new to Norwich Thorpe in May 1951 No. 70012 *John of Gaunt* is seen hard at work on Haughley bank. *(Barry Richardson)*

Middle left. Passing Haughley station, former junction for the Mid Suffolk Light Railway, is another of Norwich Thorpe's new Britannia Pacifics No. 70008 *Black Prince* heading "The Broadsman" in this undated view of the down service. *(Barry Richardson)*

Below. "The Easterling" slows to 25 mph as it enters Ipswich hauled by B17/1 class No. 61647 *Helmingham Hall* Operating between 1950-58 the service ran from Liverpool Street to Lowestoft with a portion for Yarmouth South Town, the train dividing at Beccles the only intermediate stop. Easterling was a rather archaic name for a native living to the east the term applying particularly to merchants from the Baltic. *(Flint and Harbart)*

In the next issue covering British Railways in 1951

FIRST OF THE STANDARD CLASSES No. 70000 *BRITANNIA* APPEARED IN JANUARY

Barely four months old No. 70000 *Britannia* is seen on Brentwood bank with "The Norfolkman" on 7 April 1951. *(Roy Vincent)*

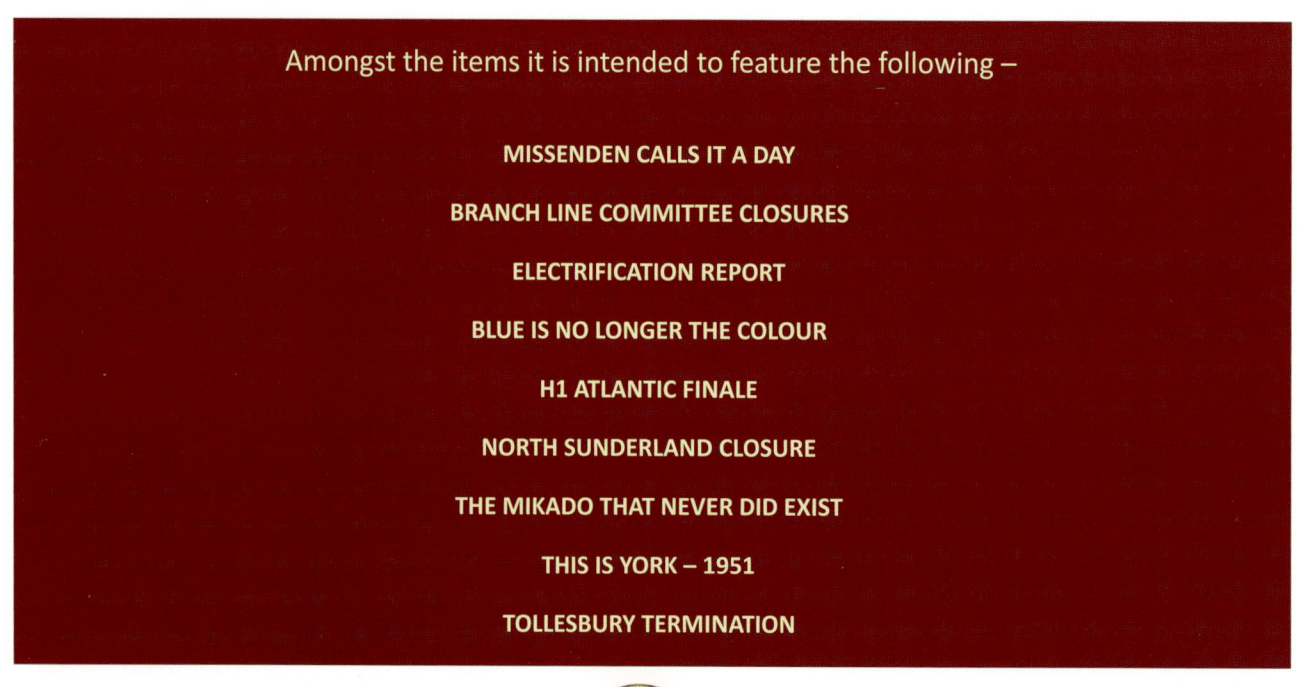

Amongst the items it is intended to feature the following –

MISSENDEN CALLS IT A DAY

BRANCH LINE COMMITTEE CLOSURES

ELECTRIFICATION REPORT

BLUE IS NO LONGER THE COLOUR

H1 ATLANTIC FINALE

NORTH SUNDERLAND CLOSURE

THE MIKADO THAT NEVER DID EXIST

THIS IS YORK – 1951

TOLLESBURY TERMINATION